Natural Resources, Environment, and Food and Energy Security Updates in North Sumatra

nova
science publishers
www.novapublishers.com

Natural Products and Therapeutics

More information about this series can be found at
https://novapublishers.com/product-category/series/natural-products-and-therapeutics/

Food Science and Technology

More information about this series can be found at
https://novapublishers.com/product-category/series/food-science-and-technology-series/

Mohammad Basyuni
Severino Salmo III
Editors

Natural Resources, Environment, and Food and Energy Security Updates in North Sumatra

nova
science publishers
www.novapublishers.com

DOI: https://doi.org/10.52305/JPRQ7444

Library of Congress Cataloging-in-Publication Data

Names: Basyuni, Mohammad editor | Salmo, Severino, III editor
Title: Natural resources, environment, and food and energy security updates in North Sumatra / Mohammad Basyuni, Severino Salmo III, Editors.
Description: Hauppauge : Nova Science Publishers, [2025] | Series: Natural products and therapeutics | Includes bibliographical references and index. |
Identifiers: LCCN 2025010284 (print) | LCCN 2025010285 (ebook) | ISBN 9798895302835 hardcover | ISBN 9798895304723 adobe pdf
Subjects: LCSH: Environmental degradation--Indonesia--Sumatera Utara | Natural resources--Indonesia--Sumatera Utara | Sustainable development--Indonesia--Sumatera Utara | Food security--Indonesia--Sumatera Utara | Energy security--Indonesia--Sumatera Utara | Sumatera Utara (Indonesia)--Environmental conditions
Classification: LCC GE160.I5 N38 2025 (print) | LCC GE160.I5 (ebook) | DDC 333.709598/12--dc23/eng/20250331
LC record available at https://lccn.loc.gov/2025010284
LC ebook record available at https://lccn.loc.gov/2025010285

Published by Nova Science Publishers, Inc. † New York

Contents

Preface

In an age where sustainable development is paramount, North Sumatra's rich tapestry of natural resources and diverse ecosystems is both a blessing and a challenge. As we confront the dual pressures of environmental degradation and the growing demand for food and energy, it becomes essential to rethink and innovate beyond traditional practices. This book, "Natural Resources, Environment, Food and Energy Security Updates in North Sumatra," aims to provide an insightful examination of these critical issues. It also aims to offer a comprehensive update on the current status and future prospects of the efforts to ensure the region's sustainability. This book is a testament to the collaborative efforts of the researchers, practitioners, and policymakers who are committed to understanding and addressing the complex interdependencies between natural resources, environmental health, and the food and energy security sectors in North Sumatra.

North Sumatra's natural resources, from its lush forests to its abundant marine life, are integral to the region's economy and the well-being of its communities. However, the traditional ways of managing these resources are increasingly proving insufficient in the face of climate change, population growth, and industrialization. To ensure the long-term security of food and energy, we must embrace innovative practices that not only protect but also enhance our natural environments. But this faces significant challenges that demand urgent and innovative solutions. The chapters within this book delve into various aspects of these challenges and offer valuable insights and practical approaches to sustainably manage and protect our natural environment.

We extend our heartfelt gratitude to all the contributors who have shared their expertise and findings. Their dedication to research and their willingness to engage in interdisciplinary dialogue have made this publication possible. We also acknowledge the support and encouragement from academic institutions, governmental agencies, and non-governmental organizations that have played a crucial role in facilitating this work.

In the realm of agriculture, the chapter explores the introduction of agroforestry and organic farming practices. These sustainable farming methods not only improve soil health and crop yields but also reduce the environmental impacts of agriculture. Agroforestry, which integrates trees and shrubs into crop and livestock systems, enhances biodiversity and provides additional sources of income for farmers. Organic farming reduces reliance on chemical inputs, leading to healthier ecosystems and communities. By transforming conventional agricultural practices, these innovations promise to enhance food security and sustainability.

Energy security is another critical focus, with an emphasis on the development of renewable energy projects. The chapter highlights successful initiatives in solar and wind energy, which are particularly relevant for remote and rural areas of North Sumatra. These renewable energy projects not only reduce greenhouse gas emissions but also provide reliable and affordable energy to communities that are often underserved by traditional power grids. The broader adoption of these technologies can catalyze economic development and improve the quality of life for many residents.

The chapter on conservation highlights the importance of community-based conservation programs. By involving local communities in conservation efforts, these programs ensure that the benefits of sustainable practices are widely shared. Empowering communities with the knowledge and tools to manage their natural resources fosters a sense of ownership and responsibility, leading to more enduring and effective conservation outcomes.

In conclusion, this book provides a detailed and nuanced exploration of the natural resources, environmental challenges, food and energy security in North Sumatra. Through the lens of innovative practices and technologies, it offers a roadmap for transforming traditional approaches into sustainable solutions that benefit both local communities and the broader region. As we look into the future, it is clear that embracing innovation and fostering collaboration are key to achieving a balanced and resilient environment for generations to come.

It is our hope that this book will serve as a valuable resource for scholars, practitioners, and policymakers alike, fostering informed decision-making and inspiring further research and collaboration. By sharing knowledge and strategies, we aim to contribute to a sustainable and resilient future for North Sumatra and beyond.

Special thanks to our reviewers: Alison KS Wee, Frida Sidik, Siti Halimah Larekeng, Andi Aznan Aznawi and Alfian Mubaraq.

Dear Readers,

As the editors of this significant publication, we are honored to present "Natural Resources, Environment, Food and Energy Security Updates in North Sumatra." This book represents a collective effort to address some of the most pressing issues of our time—issues that are particularly relevant to the region of North Sumatra, Indonesia.

The interconnectedness of natural resources, environmental sustainability, and the pillars of food and energy security cannot be overstated. Each chapter in this book provides a critical analysis of these interconnections, supported by empirical data and case studies from North Sumatra. The insights offered here are intended not only to enhance our understanding but also to guide practical actions that can lead to sustainable development.

We are deeply grateful to the authors who have contributed their time and knowledge to this endeavor. Their work reflects a diverse range of perspectives and disciplines, united by a common goal of fostering sustainable practices and policies. We also appreciate the support from various institutions and organizations that have made this publication possible.

As you engage with the content of this book, we hope you find it both informative and inspiring. It is our aspiration that this volume will stimulate further research, dialogue, and action towards achieving a balanced and sustainable future for North Sumatra.

Thank you for your interest and commitment to these critical issues.

Sincerely,

Mohammad Basyuni
and Severino Salmo III
Editors

Chapter 1

A Systematic Review of the Current Status and Significance of the Dryland and Wetland Forests in North Sumatra

Mohammad Basyuni[1,2,*]
and Etti Sartina Siregar[1,3]

[1]Center of Excellence for Mangrove, Universitas Sumatera Utara, Medan, Indonesia
[2]Department of Forestry, Faculty of Forestry, Universitas Sumatera Utara, Medan, Indonesia
[3]Department of Biology, Faculty of Mathematics and Natural Sciences, Universitas Sumatera Utara, Medan, Indonesia

Abstract

North Sumatra, Indonesia, exhibits a diverse range of ecosystems, including drylands and wetlands (mangroves and peatlands). These two ecosystem types play crucial roles in the environmental and ecological balance of the region. A systematic review of the current status and significance of dryland and wetland forests in North Sumatra was conducted using Scopus and Google Scholar databases. A total of 175 studies were derived from 125 Scopus datasets and 50 from the Google Scholar datasets. Abstracts and titles were first screened with questions of whether the studies were conducted in North Sumtra, whether they focused on forests (dryland or wetland), and whether they were assessments of biodiversity or distribution and assessments of forest conservation. All the studies that passed all criteria went on to the next step of full paper filtering; 36 studies were included in the review. Ongoing challenges discussed in this review include deforestation, land

[*] Corresponding Author's Email: m.basyuni@usu.ac.id.

In: Natural Resources, Environment, and Food and Energy Security …
Editors: Mohammad Basyuni and Severino Salmo III
ISBN: 979-8-89530-283-5

conversion, climate change, and conservation efforts. This systematic review provides a comprehensive overview of the distribution and biodiversity of the drylands and wetlands in northern Sumatra. This chapter highlights the significance of the dryland and wetland forests in North Sumatra and informs policy, conservation, and future research directions.

Keywords: conservation, restoration, wetlands, dry land, land use change, mangrove, terrestrial, North Sumatra

Introduction

North Sumatra, Indonesia, is home to diverse and unique ecosystems, including extensive dryland and wetland forests that play a crucial role in supporting biodiversity and providing ecosystem services. These forests are characterized by high levels of endemism and are vital for the survival of several plant and animal species (Laumonier et al. 2010; Basyuni et al. 2018a).

However, these forests are under increasing pressure from human activities, including deforestation, illegal logging, agricultural expansion, and infrastructure development (Basyuni et al. 2018a). These activities have led to habitat loss, fragmentation, and degradation, threatening the biodiversity and ecological integrity of these forests (Basyuni et al. 2018b; Rahmawaty et al. 2022).

Despite their ecological significance, there is a lack of comprehensive and up-to-date information on the status and significance of dryland and wetland forests in northern Sumatra. Existing studies are often fragmented and focus on specific aspects of these forests, making it challenging to develop holistic conservation strategies. A systematic review of the current status and significance of these forests is essential to fill knowledge gaps, synthesize existing information, and provide a comprehensive overview of their ecological importance, threats, and conservation needs. This work helps guide conservation efforts, inform land-use planning, and promote sustainable management practices for valuable ecosystems in North Sumatra. This chapter aims to synthesize the existing knowledge on these forests, highlighting their biodiversity, ecosystem services, threats, conservation efforts, and socioeconomic importance.

By consolidating the available information and identifying knowledge gaps, this chapter seeks to provide valuable insights for policymakers, conservation practitioners, and researchers working towards the conservation and sustainable management of North Sumatra's forests for future generations.

Methods

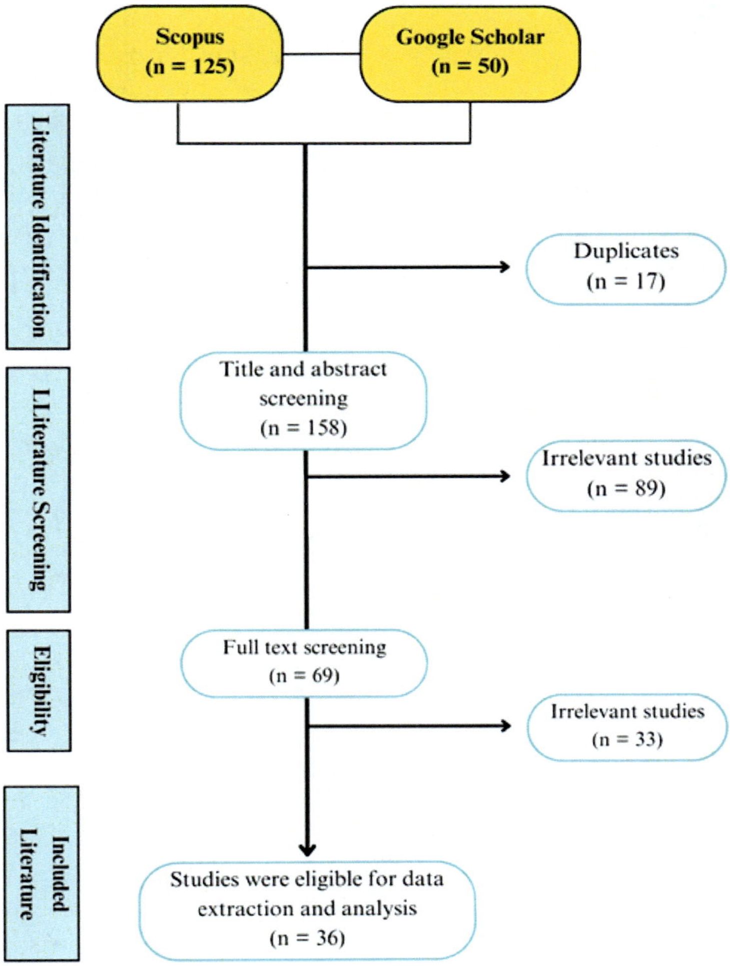

Figure 1. The PRISMA diagram showing the systematic review flow in this paper.

The primary review question is as follows: What literature and data are available on drylands and wetlands and future research in North Sumatra? This systematic review focused on studies that have reported forest types (dryland and wetland), biodiversity, and composition in North Sumatra. Data were collected from the published literature obtained through searches of preselected bibliographic databases, including Scopus and Google Scholar (Figure 1) as previously described (Illian et al. 2021; Sasmito et al. 2023; Basyuni et al. 2024). Abstracts and titles were first screened with these questions of (1) whether the studies were conducted in North Sumatra, (2) whether they focused on forests (dryland or wetland), and (3) whether they were assessments of biodiversity or distribution and assessments of forest conservation. All the studies that passed all criteria went on to the next step of full paper filtering; with the questions of (1) whether the study report distribution and diversity of forest, (2) whether the study focus on dryland or wetland, (3) whether the study present any dataset related to deforestation, (4) whether the study present assessment on forest conversion, studies are included at least two full texts screening were decided for next assessment. Finally, 36 studies were included in the review.

The population, intervention, comparator, and outcome (PICO) of the review were initially defined using an internationally accepted standard approach to systematic reviews (Illian et al. 2021; Sasmito et al. 2023; Basyuni et al. 2024). Most methodological steps in this systematic review followed previous systematic reviews in environmental science (Mengist et al. 2020; Basyuni et al. 2024; Mubaraq et al. 2024) and guided systematic reviewers within this scientific discipline (Illian et al. 2021).

Results

The number of publications on drylands and wetlands in North Sumatra increased from one in 2010-2015 to eight in 2018, then decreased to two (2020), and increased to (11) in 2022 (Table 1). There were publications (no more than two) published between 1993 and 2013. The number of papers published increased from one in 2014 to 4 in 2021, then increased to 11 in 2022, and then decreased to 3 in 2023 (Table 1). There are two types of recent publications - journal articles and conference proceedings. Journal articles are the most popular, followed by conference and review articles.

Table 1. The trend of publication on dryland and wetland forests in North Sumatra: (a) year vs the number of publications; (b) types of publication reported 2010-2023 (dated, 3 February 2024)

Year	Number of publications
2010	1
2014	1
2015	1
2018	8
2019	5
2020	2
2021	4
2022	11
2023	3
Types of publication journal article	Publication reported 26
proceeding	10

Table 2. The main data, forest type, and sites in North Sumatra from 36 studies

Data presented	Forest type	Site (North Sumatra)	Reference
Deforestation and reforestation 1990–2015	Dryland and wetland	33 regencies/city	Basyuni et al. 2018a
Mixture forest of Dipterocarpaceae.	Dryland	Langkat	Basyuni et al. 2019a
Increase and decrease in land cover	Wetland	Deli Serdang	Rahmawaty et al. 2021
10 types of land cover change	Dryland	Deli Serdang	Kurniawan et al. 2022
Trends in land use areas	Dryland	Deli Serdang	Verawaty and Firmansyah 2023.
Land cover changes 2000–2019	Dryland	Langkat	Thoha et al. 2022
12 types of land cover changes	Dryland	Langkat	Rahmawaty et al. 2022
Deforestation of mangrove forests	Wetland	Medan, Langkat, Deli serdang	Ginting et al. 2022
Mangrove species diversity in forest restoration	Wetland	Deli Serdang	Onrizal et al. 2019a
Vertical structure of mangrove forests	Wetland	Deli Serdang	Onrizal, et al. 2019b.
Floristic and carbon stock estimation	Wetland	Deli Serdang	Harefa et al. 2023
Mangroves and importance index	Wetland	Deli Serdang	Muhtadi et al. 2020
DNA amplification of mangrove plants	Wetland	Deli Serdang	Syahbana et al. 2021
Effect growth of mangrove species	Wetland	Langkat	Basyuni et al. 20218b

Table 2. (Continued)

Data presented	Forest type	Site (North Sumatra)	Reference
Reforestation mangrove	Wetland	Langkat	Basyuni et al. 2019
Land use and land cover to evaluate mangrove forest	Wetland	Langkat	Slamet et al. 2019
Ecotourism and development of mangrove forest	Wetland	Langkat and Serdang Bedagai	Basyuni, 2018
Carbon stock and rehabilitation	Wetland	Langkat	Basyuni and Simanjutak 2021
Copper and lead on *R. apiculata*	Wetland	Langkat	Leidonald et al. 2022
Salinity effect on mangrove growth and total carbon	Wetland	Langkat	Basyuni et al. 2018
Structure, carbon, and sedimentation on mangrove forest	Wetland	Langkat and Deli Serdang	Hanggara et al. 2021
Nutritional food from mangroves	Wetland	Langkat	Nawar et al. 2022
Seedling growth and carbon on the root	Wetland	Langkat	Basyuni et al. 2020
Land use and land cover for rehabilitation	Wetland	Deli Serdang	Rahmawaty et al. 2022
Mapping and analysis of land-use changes	Wetland	Langkat	Basyuni et al. 2018c.
Rehabilitation and impact to socio economic community	Wetland	Langkat	Basyuni et al. 20218
Physicochemical of macrozoobenthos	Wetland	Langkat	Basyuni et al. 2018
Hydrological data on mangrove status	Wetland	Medan, Deli Serdang, and Langkat	Balke et al. 2021
Diversity of mangrove-associated zonation	Wetland	Langkat	Nawar et al. 2021
Macrozoobenthic assemblage as restoration indicator	Wetland	Langkat, Deli Serdang	Basyuni et al. 2022
Mangrove rehabilitation on ex ponds	Wetland	Langkat	Amelia et al. 2023
Mangrove rehabilitation using *R. mucronata*	Wetland	Langkat	Basyuni et al. 2018
Oil palm and aquaculture sources of deforestation and GHG	wetland	Langkat and Deli Serdang	Basyuni et al. 2015
Salt tolerance of the mangrove seedlings	Wetland	Medan	Basyuni et al. 2014
Eco-floristic sectors stratification	Dryland	Asahan	Laumonier et al. 2010
Physicochemical primary production of mangrove litter	Wetland	Langkat, Deli Serdang	Basyuni et al. 2021
Species diversity and carbon stock	Wetland	Deli Serdang	Harefa et al. 2022

Table 2 presents the main findings in terms of the current status of drylands (6) and wetlands (31), including site studies and data. Langkat presented the highest number of study sites (25), followed by Deli Serdang

(17), Medan City (four), and Asahan (two). Studies on wetlands have primarily focused on mangrove forests (Table 2).

Discussion

By synthesizing existing knowledge, this chapter highlights several key points. Dryland and wetland forests in North Sumatra are characterized by high levels of biodiversity, with several species of flora and fauna, providing habitats for a wide range of endangered and endemic species. These forests provide vital ecosystem services, such as carbon sequestration, water regulation, and soil conservation. They also support local communities by providing resources for livelihoods, such as food, medicine, and building materials.

Despite their ecological importance, the dryland and wetland forests in North Sumatra face numerous threats, including deforestation, illegal logging, land conversion, and infrastructure development. These activities have led to habitat loss, fragmentation, and degradation, posing significant risks to biodiversity and ecosystem functions.

Several conservation efforts are in place to protect these forests, including the establishment of protected areas, community-based conservation initiatives, and sustainable land-use practices. However, these efforts face challenges, such as limited funding, inadequate enforcement, and conflicting land-use interests.

In terms of knowledge gaps, the review identifies several knowledge gaps, including the need for more comprehensive biodiversity surveys, long-term monitoring studies, and assessments of the effectiveness of conservation interventions. Addressing these gaps is crucial for improving the current understanding of these forests and for guiding future conservation actions.

Conclusion

This systematic review highlights the importance of dryland and wetland forests in North Sumatra and the need for urgent conservation efforts to protect these valuable ecosystems. By incorporating the findings of this chapter into conservation planning and policy development, stakeholders can collaborate to ensure the long-term sustainability of forests for future generations. This study provides the importance of dryland and wetland forests in North Sumatra and leads policy, conservation, and potential research initiatives.

Acknowledgments

The authors are grateful to Alfian Mubaraq, Aznan Azanwi and Shofiah Sabilah Al Mustaniroh (Universitas Sumatera Utara) for assistance in collecting published studies from Scopus and Google Scholar databases. This research is funded by Center of Excellence for Mangrove grants from Ministry of Education, Culture, Research and Technology and Penelitian Kolaborasi Unggulan Universitas Sumatera Utara (PKUU) EQUITY 2023 to Center of Excellence for Mangrove 2023-2024 (91/UN5.2.3.1/PPM/KPEP/2023).

Disclaimer

None.

References

Amelia, R., Basyuni, M., Alfinsyahri, A., Sulistiyono, N., Slamet, B., Bimantara, Y., Arifanti, V. B. 2023. Evaluation of Plant Growth and Potential of Carbon Storage in the Restored Mangrove of an Abandoned Pond in Lubuk Kertang, North Sumatra, Indonesia. *Forests* 14(1): 158.

Balke, T., Vovides, A., Schwarz, C., Chmura, G. L., Ladd, C., Basyuni, M. 2021. Monitoring tidal hydrology in coastal wetlands with the "Mini Buoy": applications for mangrove restoration. *Hydrology and Earth System Sciences*, 25(3), 1229-1244.

Basyuni, M., Puspita, M., Rahmania, R., Albasri, H., Pratama, I., Purbani, D., Kajita, T. 2024. Current biodiversity status, distribution, and prospects of seaweed in Indonesia: A systematic review *Heliyon* 10(10): E31073.

Basyuni, M., Bimantara, Y., Cuc, N. T., Balke, T., Vovides, A. G. 2022. Macrozoobenthic community assemblage as key indicator for mangrove restoration success in North Sumatra and Aceh, Indonesia. *Restoration Ecology* 30(7):e13614.

Basyuni, M., Simanjutak, E. O. 2021. Species composition and carbon stock estimation in Pulau Sembilan secondary mangrove forests, North Sumatra, Indonesia. *IOP Conference Series: Earth and Environmental Science* 713 (1): 012014.

Basyuni, M., Slamet, B., Sulistiyono, N., Munir, E., Vovides, A., Bunting, P. 2021. Physicochemical characteristics, nutrients, and fish production in different types of mangrove forests in North Sumatra and the Aceh Provinces of Indonesia. *Kuwait Journal of Science* 48(3):1-14

Basyuni, M., Slamet, B., Sulistiyono, N., Thoha, A. S., Bimantara, Y., Widjaja, E. A. 2019a. Species composition and plant diversity of logged-over forest in Sikundur, Gunung Leuser National Park, North Sumatra. *IOP Conference* Series*: Earth and Environmental Science* 374 (1): 012051.

Basyuni, M., Sulistiyono, N., Wati, R., Hayati, R. 2018a, February. Deforestation trend in North Sumatra over 1990-2015. *In IOP Conference Series: Earth and Environmental Science* 122: 012059.

Basyuni, M., Fitri, A., Harahap, Z. A. 2018b. Mapping and analysis land-use and land-cover changes during 1996-2016 in Lubuk Kertang mangrove forest, North Sumatra, Indonesia. *In IOP Conference Series: Earth and Environmental Science* (Vol. 126, p. 012110).

Basyuni, M., Harahap, M. A., Wati, R., Slmaet, B., Thoha, A. S., Nuryawan, A. & Yusriani, E. 2018. Evaluation of mangrove reforestation and the impact to socioeconomic-cultural of community in Lubuk Kertang village, North Sumatra. *In IOP Conference Series: Earth and Environmental Science* (Vol. 126, No. 1, p. 012113).

Basyuni, M., Keliat, D. A., Lubis, M. U., Manalu, N. B., Syuhada, A., Wati, R., & Yunasfi. 2018. Growth and root development of four mangrove seedlings under varying salinity. *In IOP Conference Series: Earth and Environmental Science* (Vol. 130, p. 012027).

Basyuni, M., Nuryawan, A., Putri, L. A. P., Baba, S. 2018. Effect of long-term salinity on the growth and biomass of two non-secretor mangrove plants Rhizophora apiculata and Ceriops tagal. *In IOP Conference Series: Earth and Environmental Science* (Vol. 122, No. 1, p. 012042).

Basyuni, M., Bimantara, Y., Siagian, M., Wati, R., Slamet, B., Sulistiyono, N. & Leidonad, R. 2018. Developing community-based mangrove management through eco-tourism in North Sumatra, Indonesia. *IOP Conference Series: Earth and Environmental Science* 126(1): 012109.

Basyuni, M., Telaumbanua, T. F. C., Wati, R., Sulistyono, N., & Putri, L. A. P. 2018. Evaluation of Rhizophora mucronata growth at first-year mangrove restoration at abandoned ponds, Langkat, North Sumatra. *In IOP conference series: earth and environmental science* 126: 012118.

Basyuni, M., Putri, L. A. P., Murni, M. B. 2015. Implication of land-use and land-cover change into carbon dioxide emissions in Karang Gading and Langkat Timur wildlife reserve, North Sumatra, Indonesia. *Jurnal Manajemen Hutan Tropika*, 21(1), 25-35.

Basyuni, M., Putri, L. A. P., Nainggolan, B., & Sihaloho, P. E. 2014. Growth and biomass in response to salinity and subsequent fresh water in mangrove seedlings Avicennia marina and Rhizophora stylosa. *Jurnal Manajemen Hutan Tropika*, 20(1): 17-25.

Ginting, Y. R. S., Komarudin, G., Carr, L. M. 2022. Study of changes in mangrove forest cover in three areas located on the east coast of North Sumatra Province between 1990 and 2020. *Journal of Tropical Forest Science*, 34(4): 467-479.

Hanggara, B. B., Murdiyarso, D., Ginting, Y. R., Widha, Y. L., Panjaitan, G. Y., & Lubis, A. A. 2021. Effects of diverse mangrove management practices on forest structure, carbon dynamics and sedimentation in North Sumatra, Indonesia. *Estuarine, Coastal and Shelf Science*, 259: 107467.

Harefa, M. S., Nasution, Z., Tuhono, E., Susilowati, A. 2023. Floristic composition and carbon stock estimation under restored mangrove area in Bagan Serdang, North Sumatra, Indonesia. *Biodiversitas Journal of Biological Diversity*, 24(4).

Harefa, M. S., Nasution, Z., Mulya, M. B., & Maksum, A. 2022. Mangrove species diversity and carbon stock in silvofishery ponds in Deli Serdang District, North Sumatra, Indonesia. *Biodiversitas Journal of Biological Diversity*, 23(2).

Illian, D. N., Hafiz, I., Meila, O., Utomo, A. R. H., Nuryawan, A., Siregar, G. A., & Basyuni, M. 2021. Current status, distribution, and future directions of natural products against colorectal cancer in Indonesia: a systematic review. *Molecules*, 26(16), 4984.

Kurniawan, H., Harahap, M. M., Rauf, A. 2022. Identification of land cover types and their changes in Kutalimbaru Sub-district, Deli Serdang District, North Sumatra, Indonesia. *In IOP Conference Series: Earth and Environmental Science* 85 (1): 012035).

Laumonier, Y., Uryu, Y., Stüwe, M., Budiman, A., Setiabudi, B., & Hadian, O. 2010. Eco-floristic sectors and deforestation threats in Sumatra: identifying new conservation area network priorities for ecosystem-based land use planning. *Biodiversity and conservation*, 19, 1153-1174.

Leidonald, R., Dalimunthe, A., Rakesya, N. 2022. *Rhizophora apiculata* on copper and lead heavy metal substances and their effect on water quality in Belawan. *In IOP conference series: earth and environmental science* 995: 012043.

Mengist, W., Soromessa, T., Legese, G. 2020. Method for conducting systematic literature review and meta-analysis for environmental science research. *MethodsX* 7: 100777.

Mubaraq, A., Sembiring, M., Widiastuti, E., Fachrial, E., Utomo, B., Turjaman, M., Basyuni, M. 2024. Diversity and distribution of nitrifying bacteria play an important role in the nitrogen cycle in mangrove sediments. *Global Journal of Environmental Science and Management* 10(4): 1-18.

Muhtadi, A., Leidonald, R., Triwibowo, K., Azmi, N. 2020. Flora Fauna Biodiversity and CSR Implementation in the Mangrove Ecosystem of Bagan Serdang Village, North Sumatra Province. *Jurnal Ilmiah Perikanan dan Kelautan*, 12(1), 59-72.

Nawar, M. K., Basyuni, M., Hanum, C., Siregar, E. S. 2022. Bioprospecting opportunities of mangrove fruits for the coastal community in lubuk kertang and pulau sembilan, North Sumatra, indonesia. *Asian Journal of Plant Sciences*, 21: 145-153.

Nawar, M. K., Basyuni, M., Hanum, C. 2021. Diversity of mangrove species associated with zonation in Lubuk Kertang Village and Pulau Sembilan, North Sumatra. *In IOP Conference Series: Earth and Environmental Science* 912: 012007.

Onrizal, O., Amelia, R., Arif, M., Agustini, R., Amri, K., Simamora, G. N., Mansor, M. (2019a, May). The role of forest restoration in conserving mangrove plant at the eastern coast of North Sumatra. *In IOP Conference Series: Earth and Environmental Science* 260: 012072

Onrizal, O., Amelia, R., Amri, K., Sulistiyono, N., Mansor, M. 2019b. Stand structure and diversity of restored mangroves at abandoned pond in eastern coast of North Sumatra. *In IOP Conference Series: Earth and Environmental Science* 305: 012050).

Rahmawaty, R., Rauf, A., Harahap, M. M., & Kurniawan, H. 2022. Land cover change impact analysis: an integration of remote sensing, GIS and DPSIR framework to deal with degraded land in Lepan Watershed, North Sumatra, Indonesia. *Biodiversitas Journal of Biological Diversity*, 23(6): 3000-3011.

Rahmawaty, Rauf, A., Harahap, M. M., Kurniawan, H. 2022. Analysis of land-use change over five-and ten-year periods in Hamparan Perak, North Sumatra, Indonesia. *Geocarto International,* 37(27), 15037-15062.

Rahmawaty, Harahap, M. M., Kurniawan, H., Mandasari, D. 2021. DPSIR1 model approach to address land-use changes in Deli Serdang District, North Sumatra Province. *In IOP Conference Series: Earth and Environmental Science* (Vol. 724, No. 1, p. 012028). IOP Publishing.

Slamet, B., Nadira, S., Basyuni, M. 2019. Land use analysis of coastal belt in the coastal area of Langkat Regency North Sumatra Province using Sentinel 2 Satellite imagery. *In IOP Conference Series: Earth and Environmental Science* 374: 012044.

Syahbana, R., Basyuni, M., Siregar, L. A. M. 2021. Isolation and amplification of mangrove plants using DNA barcode in Percut Sei Tuan, North Sumatra, Indonesia. *In IOP Conference Series: Earth and Environmental Science* (Vol. 912, No. 1, p. 012028).

Thoha, A. S., Saraswita, N., Sulistiyono, N., Wiranata, D., Sirait, S. M., Inaldi, R. 2022. Analysis of land cover changes due to forest fires in Gunung Leuser National Park, North Sumatra Province, Indonesia. *Biodiversitas Journal of Biological Diversity*, 23(3): 1420-1426.

Verawaty, I., Firmansyah, I. 2023. Modeling of land use and cover changes (LUCC) in Deli Serdang Regency, North Sumatra Province. *Jurnal Pengelolaan Sumberdaya Alam dan Lingkungan (Journal of Natural Resources and Environmental Management)*, 13(2): 237-251.

Chapter 2

The Implications of Land Use and Land Cover Changes on the Natural Environment in North Sumatra

Rahmawaty*
and Delvian
Faculty of Forestry, Universitas Sumatera Utara, Pancur Batu, Deli Serdang, Indonesia
Natural Resources and Environmental Management Study Program, Postgraduate School, Universitas Sumatera Utara, Indonesia

Abstract

The issue of Land Use and Land Cover Changes (LULCCs) is closely related to population growth in a region, including North Sumatra. It is crucial to consider LULCCs as they have implications for North Sumatra's natural resources and the environment. This study analyzed and assessed changes in land cover and land use in North Sumatra and their implications for natural resources and the environment. The LULCC analysis employed remote sensing technology and the Geographic Information System (GIS) to understand the dynamics of land changes over a period. Priority factors causing land cover changes were analyzed using Analytical Hierarchy Process (AHP). Literature studies and interviews with key informants related to LULCCs were conducted in various areas of North Sumatra. The study's findings indicate that changes in land cover and land use over time had significant impacts on natural resources and the environment. Generally, economic factors were the primary triggers for land use changes. Policy interventions are needed to address these changes and

* Corresponding Author's Email: rahmawaty@usu.ac.id.

In: Natural Resources, Environment, and Food and Energy Security …
Editors: Mohammad Basyuni and Severino Salmo III
ISBN: 979-8-89530-283-5

to ensure the sustainability of ecosystems and controlled reduction of natural vegetation. The preservation of forests is essential to prevent a decrease in forest cover, which could lead to the loss of natural habitats and biodiversity. Additionally, the existence of endemic flora and fauna populations must be maintained. Preserving forests is crucial for maintaining the quality of soil and water, which provide essential natural resources for human life. LULCC is identified as an approach to gain a better understanding of conservation challenges and natural resource management, especially in North Sumatra. Therefore, the insights from this study are expected to contribute significantly to the context of understanding LULCCs and provide a foundation for sustainable policy decision-making in North Sumatra.

Keywords: conservation, climate, ecosystem, forest, SDGs, North Sumatra

Introduction

North Sumatra Province is located in the western part of Indonesia at 1°–4° North Latitude and 98°–100° East Longitude and spans an area of approximately 72,460.744 square kilometers. The territorial boundaries of North Sumatra Province are as follows: north bordering Aceh Province, west bordering the Indian Ocean, south bordering the Riau and West Sumatra Provinces, and east bordering the Strait of Malacca (Figure 1).

North Sumatra has a diverse population comprising various ethnic groups, including the Batak, Chinese, Javanese, and Malay. The capital and largest city is Medan, which serves as a major economic and transportation hub in the region. Population density varies across different parts of the province, with urban areas experiencing higher population concentrations than the rural hinterlands. North Sumatra's economy is predominantly agrarian, with agriculture, forestry, and fisheries as its primary economic activities. The province is renowned for its production of palm oil, rubber, coffee, and various agricultural commodities that contribute significantly to Indonesia's export earnings. It is famous for its rich biodiversity, diverse landscapes, and significant socioeconomic activities. However, like many regions worldwide, North Sumatra is undergoing rapid land use and land cover changes (LULCCs), driven primarily by population growth, urbanization, agricultural expansion, and industrial development.

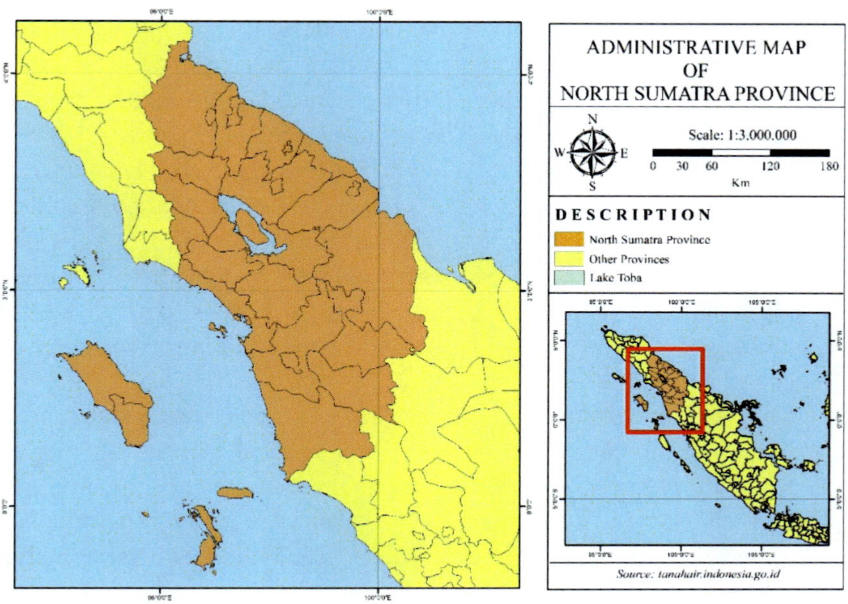

Figure 1. Administration map of North Sumatra Province.

Some land covers in North Sumatra, according to the Ministry of Environment and Forestry in 2023, include: primary dry land forest, secondary dryland forest, forest plantation, plantation, shrubs, dryland agriculture, mixed dryland agriculture, paddy field, settlement, open field, grass, water body, primary mangrove forest, secondary mangrove forest, primary swamp forest, secondary swamp forest, swamp thicket, pond, airport, mining, and swamp. Previous studies conducted in North Sumatra, indicated that several changes in land cover and land use occurred, such as changes in forest cover to plantations, settlements, agricultural land, and industries (Bambang, Sanudin, and Sukmana 2008; Basyuni, Putri, and Murni 2015, 2018; Daulay et al. 2023; Kuswanda and Sunandar 2019; Rahmawaty et al. 2020–2023; Saputra and Lee 2019; Verawaty, Widiatmaka, and Firmansyah 2023; Wahyuni, Guchi, and Hidayat 2014). This shows that North Sumatra is experiencing significant transformations in land cover and land use patterns, reflecting broader global trends of urbanization, industrialization, and agricultural expansion. These changes can have both positive and negative impacts on the environment, biodiversity, and local communities.

On the positive side, urbanization and industrial development can bring economic opportunities, job creation, and improved infrastructure to the region, potentially raising residents' living standards. Agricultural expansion can boost food production and contribute to economic growth. However, these changes often come at a cost. Deforestation of plantations and settlements can lead to habitat loss, biodiversity decline, and environmental degradation (Daulay et al. 2023; Pasaribu et al. 2023). Agricultural expansion may also strain natural resources, such as water and soil, leading to long-term sustainability challenges (Nugroho et al. 2013). In addition, rapid urbanization can result in issues such as pollution, congestion, and inadequate housing.

Understanding the dynamics of land use and land cover change is crucial for policymakers, planners, and local communities to mitigate negative impacts and promote sustainable development in North Sumatra. This requires an integrated approach that balances economic development, environmental conservation, and social equity. Collaborative efforts involving government agencies, research institutions, civil society organizations, and local communities are essential for effective land use planning and management strategies. This study aimed to review the implications of LULCCs on natural resources and the environment in North Sumatra.

Literature Review of Land Use Land Cover Changes and Their Implications on Natural Resources and Environment

Historical Trends in Land Cover and Land Use Changes

Historically, North Sumatra has experienced significant deforestation and the conversion of forests into agricultural land, particularly for palm oil and rubber plantations, such as those in the Deli Serdang and Langkat Regencies (Basyuni, Putri, and Murni 2015; Kuswanda and Sunandar 2019; Rahmawaty et al. 2019, 2020, 2022). This expansion of agricultural land has been driven by population growth, government policies promotion of agricultural development, and the increasing global demand for commodities. Consequently, large swathes of primary forests have been cleared, leading to habitat loss, soil degradation, and biodiversity decline. Urbanization has also been a notable trend in North Sumatra, especially in and around major cities, such as Medan (Rahmawaty et al. 2023). Rapid population growth, rural-to-urban migration, and economic opportunities have fueled urban expansion, leading to the conversion of rural land for residential, commercial, and industrial purposes. This urban sprawl has implications for land use patterns, natural resource management, and environmental sustainability, as it often encroaches on agricultural land, green spaces, and sensitive ecosystems. Previous research has been conducted on the land use/land cover in North Sumatra, as listed in Table 1.

Table 1. Summary of some research related to land use/land cover in North Sumatra

Title	Authors, Year	Key findings
Changes in Forest Functions in Asahan District, North Sumatra	Bambang et al., 2008	The condition of the forest in Asahan Regency, North Sumatra including protected forests, limited production forests, production forests, and conversion forests, have partially changed their function to other land uses such as settlements, plantations, rice fields, ponds, and so on.
Prediction of land use and land cover changes for North Sumatra, Indonesia, using an artificial-neural-network-based cellular automaton	Saputra et al., 2019	The Land use and land cover (LULC) projections for 2050 and 2070 indicate significant expansion in plantation areas by more than 4%. Conversely, forest and crop areas are expected to decline by approximately 1.2% and 1.6% respectively, by 2050, and by 1.2% and 1.7% respectively, by 2070. These trends suggest human-induced shifts in LULC, with forests and croplands being converted into plantations.
Implication of land-use and land-cover change into carbon dioxide emissions in Karang Gading and Langkat Timur wildlife Reserve, North Sumatra, Indonesia	Basyuni et al., 2015	A land-use change matrix revealed that the decline in mangrove forest area war attributed to the expansion of other land uses, notably aquaculture (50.00%) and oil palm plantation (28.83%)
Carbon dioxide emissions from forestry and peat Land using land-use/land-cover changes in North Sumatra, Indonesia	Basyuni et al., 2018a	The findings revealed considerable increases in emission from intensive land uses in this province during the historical period (2006-2012), notably in dry land agriculture (109.6%), paddy fields (16.23%), and estate plantations (15.11%). Conversely, significant decreases were observed in forest cover, including secondary dry land forests (7.60%), secondary mangrove forest (9.03%), and secondary swamp forests (33.98%). The most substantial reduction occurred in mixed dry land agriculture 79.96%)
Deforestation trend in North Sumatra over 1990-2015	Basyuni et al., 2018b	The findings revealed that out of the 33 regencies/cities in North Sumatera, 25 district experienced deforestation, with Labuhanbatu and South Labuhanbatu exhibiting the highest rates (2,238.08 and 1,652.55 hectares per year, respectively). Only one area underwent deforestation, while seven districts demonstrated neither deforestation or reforestation. Over the 25-your observation period, forest cover decreased by 22.92%, while non-forest land use increased by 11.33%. The significant loss of forest in North Sumatra underscores the importance of conservation efforts and the implementation of sustainable forest management practices.
Mapping and analysis land-use and land cover changes during 1996-2016 in Lubuk Kertang mangrove forest, North Sumatra, Indonesia	Basyuni et al., 2018c	A land use change matrix revealed a decline in mangrove forest coverage, with a loss of 109.4 hectares from 1996 to 2006, primarily converted to agriculture covering 515 hectares (47.1%). Conversely, between 2006 and 2016, mangrove loss amounted to 291.2 hectares, with the majority of deforestation occurring due to land conversion for agriculture, aquaculture, and urban expansion.

The natural resources of the province, including minerals, timber, and fisheries, have been extracted and exploited, contributing to changes in land use and land cover. In particular, mining activities have led to forest clearance, soil erosion, and pollution of waterways, thereby impacting ecosystems and local communities.

Based on Table 1, it is evident that several studies related to land use and cover in North Sumatra reported forest conversion into plantations, settlements, and other types of land cover (Table 1). This demonstrates a trend of changing land use from forests to more intensive forms of land use, such as plantations and settlements. These changes can have a significant impact on forest ecosystems, biodiversity, and the availability of natural resources in the region.

Dynamics of Land Use and Land Cover Changes in North Sumatra

LULCCs in North Sumatra affect natural ecosystems, including tropical rainforests (Bambang, Sanudin, and Sukmana 2008; Pandiangan et al. 2017), mangroves (Basyuni 2018c; Ginting et al. 2022), and wetlands. Deforestation for agriculture, logging, and infrastructure development leads to habitat loss, soil erosion, and reduced biodiversity (Basyuni, Fitri, and Harahap 2018; Basunyi et al. 2018; Rahmawaty et al. 2021–2023). Moreover, changes in land cover influence local climate patterns, water cycles, and carbon sequestration (Lyu et al. 2023; Yu and Song 2023), exacerbating environmental challenges such as climate change and natural disasters. North Sumatra is home to a remarkable array of flora and fauna (Rahmawaty et al. 2015), including endangered species such as the Sumatran orangutan (Kuswanda et al. 2018), Sumatran tiger (Patana et al. 2022), and Sumatran elephant (Kuswanda and Sunandar 2019). Unchecked LULCCs threaten their habitats and survival. Monitoring and understanding LULCC trends are essential for identifying priority areas for conservation, implementing habitat restoration initiatives, and mitigating human-wildlife conflicts. Agricultural and plantation expansion, urbanization, industrialization, and infrastructure development are major drivers of change in North Sumatra. These drivers interact in complex ways, shaping land-use patterns, socioeconomic dynamics, and environmental conditions in the region. Understanding the implications of these drivers is essential for sustainable land use planning, natural resource management, and socioeconomic development in North Sumatra.

Significance of LULCCs in North Sumatra

North Sumatra's economy relies heavily on agriculture, leading to extensive land use changes, including deforestation and the conversion of forests into agricultural land. This expansion poses challenges, such as habitat loss, soil degradation, and biodiversity decline, while also impacting local communities and indigenous groups reliant on forest resources.

Rapid urbanization in cities such as Medan has resulted in the conversion of rural land for residential, commercial, and industrial purposes. Urban sprawl contributes to land fragmentation, loss of green spaces, and environmental degradation, thereby affecting water quality, air pollution, and overall urban livability. North Sumatra's rich natural resources, including forests, rivers, and marine ecosystems, are under pressure due to LULCCs. Sustainable management of these resources is critical for ensuring long-term environmental sustainability, biodiversity conservation, and socioeconomic development. LULCCs influence North Sumatra's vulnerability to climate change. Some impacts of climate change are floods (Schirpke et al. 2023; Villarreal-Rosas et al. 2022), droughts, and landslides. Sustainable land-use practices, including reforestation, watershed management, and coastal protection, are essential for enhancing resilience and adaptation to climate-related risks.

Discussion of Land Use and Land Cover Changes (LULCCs) and Their Implications

The Consequences of LULCCs on Ecosystem Services and Sustainability

The primary consequence of LULCCs on ecosystem services and sustainability in North Sumatra is their impact on natural resources, including deforestation and forest degradation, loss of biodiversity and habitat fragmentation, soil erosion and degradation, water resource depletion, and pollution. LULCCs, particularly deforestation for agricultural expansion and logging, lead to the loss of forest cover in North Sumatra. Deforestation reduces the availability of forest ecosystem services such as carbon sequestration, climate regulation, and soil stabilization (Basyuni et al. 2018; Qiu, Hu, and Zhang 2021). Forest degradation, often caused by selective logging, fire, and unsustainable land management practices, further diminishes the ecological integrity of forests and compromises its ability to provide essential services and support biodiversity. The conversion of forests and natural habitats into agricultural land, urban areas, and infrastructure has resulted in the loss of biodiversity in North Sumatra. Many species, including endemic and threatened species such as Sumatran orangutan and Sumatran tiger, depend on forest habitats for survival. LULCCs disrupt their habitats, leading to population decline and increased risk of extinction. Habitat fragmentation, caused by the fragmentation of natural landscapes into smaller, isolated patches, exacerbates biodiversity loss by restricting species movement, gene flow, and ecosystem connectivity (Rodríguez-Echeverry et al. 2018).

LULCCs contribute to soil erosion and degradation in North Sumatra, particularly in areas where forests have been cleared for agricultural or infrastructure development. Deforestation exposes the soil to erosion by wind and water, leading to the loss of topsoil, nutrient depletion, and reduced soil fertility. Unsustainable land management practices, such as intensive farming, monoculture plantations, and improper land clearance techniques accelerate soil erosion and degradation and threaten agricultural productivity and livelihoods.

LULCCs affect water resources in North Sumatra by altering hydrological cycles, reducing water availability, and increasing water pollution (Dwarakish and Ganasri 2013). Deforestation disrupts the natural water balance, leading to changes in rainfall patterns, runoff rates, and groundwater recharge. Agricultural runoff, urban sewage, and industrial effluents contribute to water pollution and contaminate rivers, lakes, and groundwater sources. This pollution has detrimental effects on aquatic ecosystems, public health, and drinking-water quality.

The Importance of Studying Land Use and Land Cover Changes (LULCCs): Linkages to Environmental Degradation and Biodiversity Loss

Understanding LULCC is crucial for several reasons, each of which has significant implications for the environment, biodiversity, and socioeconomic development. LULCCs often lead to environmental degradation, including deforestation, soil erosion, and habitat destruction. For example, when forests are cleared for agricultural purposes or urban development, it disrupts ecosystems, reduces carbon sequestration, and increases the risk of soil erosion and flooding. Studying LULCCs helps to identify areas prone to degradation and facilitates the implementation of measures to mitigate their effects, such as reforestation programs or land-use planning regulations. Changes in land use and land cover can directly affect biodiversity by altering habitats and fragmenting ecosystems. Habitat loss due to urbanization, agricultural expansion, or infrastructure development can lead to the displacement or extinction of species. By studying LULCCs, researchers can assess the extent of habitat loss, identify key biodiversity hotspots, and develop conservation strategies to protect vulnerable species and ecosystems.

LULCC and Implications for the Environment

The implications of LULCCs for the environment in North Sumatra are centered on their impacts on climate change, hydrological cycles, and vulnerability to natural hazards. LULCCs, particularly deforestation and forest degradation, contribute to climate change by releasing carbon dioxide (CO_2) and other greenhouse gases (GHGs) into the atmosphere. Forests act as carbon sinks that absorb CO_2 from the atmosphere through photosynthesis. When forests are cleared or degraded, this carbon is released back into the atmosphere, which exacerbates global warming and climate change. The conversion of forests into agricultural land, urban areas, and industrial zones in North Sumatra has resulted in significant carbon emissions, making this region a net contributor to GHG emissions. Climate change, in turn, has cascading effects on ecosystems, weather patterns, and agricultural productivity, further exacerbating environmental challenges in the region (Hasan et al. 2020).

LULCCs alter hydrological cycles by modifying the patterns of precipitation, evaporation, and water runoff in North Sumatra. Deforestation disrupts the natural water balance by reducing transpiration, interception, and soil infiltration rates in forested areas. As a result, the surface runoff increases, leading to more frequent and severe flooding during the rainy season. Conversely, decreased groundwater recharge and streamflow may lead to water scarcity and droughts in some areas during the dry season. Changes in land cover, such as the expansion of impermeable surfaces in urban areas and the conversion of natural wetlands into agricultural land, compound changes in hydrological cycles and contribute to water quality degradation and ecosystem disturbance. LULCCs increase vulnerability to natural hazards, such as floods, landslides, and wildfires (Dwarakish and Ganasri 2015).

Deforestation and land conversion reduce the natural protective functions of forests such as soil stabilization, slope reinforcement, and flood mitigation (Nugroho et al. 2013). Without the buffering capacity of forests, areas prone to erosion, landslides, and flash floods become more susceptible to natural hazards, leading to increased risks to human lives, infrastructure, and livelihoods. Additionally, changes in land cover and land use patterns may exacerbate the intensity and frequency of wildfires, particularly during periods of prolonged drought and dry conditions, further threatening ecosystems and communities (Schirpke et al. 2023).

Examination of Environmental Impacts Resulting from LULCCs

The examination of the environmental impacts of LULCCs in North Sumatra reveals significant challenges related to climate change, hydrological cycles, and natural hazards (Nugroho et al. 2013; Pan et al. 2021). Addressing these impacts requires integrated approaches that prioritize sustainable land management, ecosystem conservation, and climate change adaptation strategies to enhance environmental resilience and promote the well-being of communities and ecosystems in the region.

LULCCs, particularly deforestation and forest degradation, significantly contribute to climate change by altering the carbon balance in the atmosphere (Kalnay and Cai 2003). Forests act as carbon sinks, absorbing CO2 during photosynthesis and storing it in the biomass and soil. When forests are cleared or disturbed, stored carbon is released back into the atmosphere as CO2, which contributes to greenhouse gas emissions. The loss of forests and natural habitats in North Sumatra exacerbates global warming and climate change by reducing the carbon sequestration capacity and increasing carbon emissions. Climate change effects associated with LULCCs include rising temperatures, changes in precipitation patterns, more frequent and severe weather events (such as storms, droughts, and heat waves), and disruptions to ecosystems and livelihoods. According to Zhang et al. (2016), climate change affects people's livelihoods.

LULCCs alter hydrological cycles by affecting the distribution, quantity, and quality of water resources in North Sumatra. Deforestation reduces evapotranspiration rates and rainfall interception, leading to increased surface runoff and decreased groundwater recharge. Changes in land cover due to urbanization and agricultural expansion further modify hydrological processes by increasing impervious surfaces, reducing infiltration rates, and altering water flow patterns. These alterations in hydrological cycles contribute to water scarcity, increased flood risks, changes in river flow regimes, and the degradation of aquatic ecosystems in North Sumatra (Basyuni, Putri, and Murni 2015).

LULCCs in North Sumatra influence the occurrence and severity of natural hazards, including floods, landslides, wildfires, and coastal erosion. Deforestation and land conversion reduce the natural protective functions of forests such as soil stabilization, slope reinforcement, and flood mitigation, thereby increasing their vulnerability to erosion, landslides, and flash floods. Changes in land cover and land use patterns may exacerbate the intensity and frequency of wildfires, particularly during periods of prolonged drought and dry conditions. Coastal areas are particularly susceptible to erosion and inundation due to LULCCs, which alters coastal ecosystems, disrupts natural sediment dynamics, and increases the vulnerability of coastal communities to storm surges and sea-level rise (Basyuni, Putri, and Murni 2015; Villarreal-Rosas et al. 2022).

Implications of LULCCs on Natural Resources and the Environment in North Sumatra

Forests play a vital role in providing ecosystem services, such as carbon sequestration, climate regulation, soil stabilization, and habitat provision in North Sumatra (Basyuni et al. 2018). Deforestation and forest degradation driven by LULCCs result in the loss of these ecosystem services, leading to climate change, soil erosion, increased vulnerability to natural hazards, and loss of biodiversity. The loss of forests also affects local communities and indigenous peoples who depend on forests for their livelihoods, including food, medicine, fuelwood, and cultural practices. Forests are essential for maintaining water quality and regulating hydrological cycles, and their degradation can lead to reduced water availability, increased sedimentation, and decreased resilience to drought and flooding. North Sumatra is renowned for its rich biodiversity, including endemic species such as the Sumatran orangutan, Sumatran tiger, and numerous plant and insect species.

LULCCs, particularly deforestation and habitat fragmentation, pose a significant threat to biodiversity by destroying natural habitats and disrupting ecological processes. The loss of biodiversity has cascading effects on ecosystem functioning, including pollination, seed dispersal, nutrient cycling, and pest control, which can affect agricultural productivity and human well-being. Furthermore, a loss of biodiversity reduces the resilience of ecosystems to environmental changes and increases the risk of species extinction, leading to irreversible consequences for the ecological balance and human societies. Soil and water are essential natural resources that support agricultural productivity, biodiversity, and human livelihood in North Sumatra. LULCCs contribute to soil erosion, soil degradation, and loss of soil fertility through deforestation, unsustainable land management practices, and the conversion of natural habitats into agricultural land. Soil erosion leads to sedimentation of water bodies, reduced soil productivity, and increased vulnerability to landslides and flooding, posing risks to agricultural sustainability and food security. Changes in land cover also affect water resources by altering hydrological cycles, reducing groundwater recharge, and increasing water pollution from agricultural runoff, urban sewage, and industrial activities.

The Importance of Forest Preservation, Biodiversity Conservation, and Soil and Water Management

Forest preservation is crucial for maintaining ecosystem services, biodiversity, and climate regulation in North Sumatra. Forests act as carbon sinks, absorbing CO2 from the atmosphere and mitigating climate change. Preserving forests helps maintain carbon sequestration capacity and reduce greenhouse gas emissions. Forests provide habitats for diverse flora and fauna, including endangered species such as Sumatran orangutans and tigers. Forest preservation is essential for biodiversity conservation and protection of the ecological integrity of ecosystems. Forests also regulate hydrological cycles, prevent soil erosion, reduce flood risk, and maintain water quality. Preserving forests helps to safeguard water resources and ensures their availability for human consumption, agriculture, and industry.

Biodiversity conservation is essential for maintaining ecosystem resilience, stability, and productivity in North Sumatra. Biodiverse ecosystems are more resilient to environmental changes and disturbances because they contain a greater variety of species with diverse ecological functions and traits. Biodiversity provides essential ecosystem services such as pollination, seed dispersal, pest control, and nutrient cycling, which are vital for agricultural productivity, food security, and human well-being. Biodiversity conservation helps preserve genetic diversity, which is essential for breeding crops and developing new varieties with traits such as disease resistance, drought tolerance, and high yields.

Soil and water management are critical for sustainable land use, agricultural productivity, and ecosystem health in North Sumatra. Soil conservation practices, such as contour plowing, terracing, and agroforestry help to prevent soil erosion, improve soil fertility, and enhance agricultural sustainability. Water management techniques include rainwater harvesting, improvements in irrigation efficiency, watershed management, water conservation support, drought resilience, and sustainable water use practices. Soil and water conservation efforts contribute to ecosystem restoration, biodiversity conservation, and climate change adaptation by maintaining soil health, water quality, and ecological balance (Hasan et al. 2020).

Assessing the resilience of ecosystems in the face of changing land-use patterns involves evaluating their capacity to withstand and recover from disturbances. The key indicators of ecosystem resilience include biodiversity, ecosystem services, habitat connectivity, and adaptive capacity. Monitoring changes in land use, biodiversity loss, soil and water quality, and ecosystem health assists in identifying vulnerable areas and prioritizing conservation and restoration efforts. Integrated approaches that combine the ecological, social, and economic dimensions are essential for building resilience and promoting sustainable development in North Sumatra.

Implementing soil and water conservation efforts involves adopting sustainable land management practices, land-use planning, and policy interventions. Soil conservation measures aim to reduce soil erosion, enhance soil fertility, and protect soil health through practices such as conservation tillage, cover cropping, and reforestation. Water conservation initiatives focus on improving water-use efficiency, reducing water waste, and protecting water resources from pollution and depletion. These include promoting water-saving technologies, watershed management, and integrated water resource management approaches. Community participation, capacity building, and stakeholder engagement are crucial for the successful implementation of soil and water conservation efforts to ensure local ownership, support, and long-term sustainability.

Policy Interventions and Conservation Strategies

The implementation of strict regulations and law enforcement mechanisms to protect the remaining forest areas from illegal logging, land clearing, and encroachment is essential. These steps should not be mere words on paper but must be a solid commitment to protect our natural resources. Another positive step is to establish and expand protected areas, national parks, and protected forests to preserve critical habitats, biodiversity centers, and ecosystem services. Additionally, there is a need for community-based forest management initiatives that empower local communities to manage and conserve forests sustainably while providing alternative livelihood opportunities. This comprehensive approach will convince the public of the strategy's effectiveness and give them confidence in the environmental conservation efforts undertaken.

Another critical step is to develop a comprehensive land-use planning framework that prioritizes sustainable land management practices, biodiversity conservation, and climate resilience. In addition, zoning regulations that designate areas for various land uses such as agriculture, urban development, conservation, and buffer zones, must be implemented. These regulations are essential for preventing land-use conflicts, minimizing environmental impacts, and ensuring the harmonious coexistence between human activities and the environment.

In addition, there needs to be financial incentives, tax breaks, and subsidies for farmers, landowners, and businesses that implement sustainable land management practices, such as agroforestry, organic farming, and reforestation. These incentives aim to support the implementation of sustainable practices and improve agricultural productivity and livelihoods, thereby instilling hope and optimism regarding the potential for positive environmental and economic changes.

It is also important to note the need to enforce strict regulations regarding the conversion of land from forests and natural habitats to agricultural, industrial, or urban uses to minimize environmental degradation and the loss of biodiversity. Environmental impact assessments (EIA) are required for large-scale development projects such as infrastructure, mining, and plantation expansion to assess the potential environmental impacts and mitigate adverse impacts.

Promoting sustainable land management practices such as agroforestry and sustainable agriculture, community-based conservation initiatives, and ecosystem restoration and rehabilitation are also important. It promotes agroforestry systems that integrate trees with food crops and livestock to increase soil fertility, biodiversity, and carbon sequestration, while providing many benefits for farmers. Encourage sustainable agricultural practices, such as conservation farming, organic farming, and integrated pest management, to reduce chemical inputs, soil erosion, and water pollution. Support community-based conservation initiatives that empower local communities to sustainably manage natural resources, conserve biodiversity, and protect ecosystems, while improving livelihoods and resilience to environmental change. Training, capacity building, and technical assistance should be provided to local communities, indigenous groups, and stakeholders to increase their knowledge and skills in sustainable land management and conservation practices. Investments should be made in ecosystem restoration and rehabilitation projects such as reforestation, afforestation, and restoration of degraded land, to increase ecosystem resilience, carbon sequestration, and biodiversity conservation. Partnerships should be fostered between government agencies, NGOs, private sector organizations, and local communities to mobilize resources, share expertise, and implement large-scale restoration initiatives. This will emphasize the integral role of local communities and stakeholders, making them feel valued and part of the collective effort.

Conclusion

The implications of LULCCs for the environment in North Sumatra are far reaching, affecting climate change, hydrological cycles, and vulnerability to natural hazards. Addressing these implications requires holistic approaches that prioritize sustainable land management, ecosystem restoration, and climate change adaptation strategies to enhance environmental resilience and promote long-term sustainability in the region.

Forest preservation, biodiversity conservation, soil and water management, and ecosystem resilience assessment are essential components of sustainable land use and environmental management in North Sumatra. By prioritizing these efforts and adopting integrated approaches, stakeholders can promote ecological resilience, protect natural resources, and enhance the well-being of communities and ecosystems in the region.

Policy interventions to mitigate the adverse impacts of LULCCs and promote sustainable land management practices in North Sumatra should include a range of actions, such as forest protection, land use planning, incentives for sustainable practices, regulation of land conversion and development, promotion of agroforestry, and sustainable land management. Policymakers can address the root causes of environmental damage by adopting integrated approaches and fostering collaboration among stakeholders.

Disclaimer

None.

References

Basyuni, M., A. Fitri., Z. A. Harahap. 2018. Mapping and Analysis Land-use and Land-cover Changes during 1996-2016 in Lubuk Kertang Mangrove Forest, North Sumatra, Indonesia. *IOP Conference Series: Earth and Environmental Science* 126(1). https://doi.org/10.1088/1755-1315/126/1/012110.

Basyuni, M., Lollie A. P. Putri, Murni, M. B. 2015. Implication of Land-use and Land-cover Change into Carbon Dioxide Emissions in Karang Gading and Langkat Timur Wildlife Reserve, North Sumatra, Indonesia. *Jurnal Manajemen Hutan Tropika* 21(1). https://doi.org/10.7226/jtfm.21.1.25.

Basyuni, M., Sulistyono, N., Slamet, B., Wati, R. 2018. Carbon Dioxide Emissions from Forestry and Peat Land Using Land-use/Land-cover Changes in North Sumatra, Indonesia. *IOP Conference Series: Earth and Environmental Science* 126(1). https://doi.org/10.1088/1755-1315/126/1/012111.

Daulay, U. A., Ahmad, A. G., Purwoko, A. 2023. Analysis of Land Cover Change due to Deforestation at Holiday Resort Nature Park, North Sumatra Province, Indonesia. *Journal of Sylva Indonesiana* 6(1). https://doi.org/10.32734/jsi.v6i01.9229.

Dwarakish, G. S., Ganasri, B. P. 2015. Impact of Land Use Change on Hydrological Systems: A Review of Current Modeling Approaches. *Cogent Geoscience* 1(1). https://doi.org/10.1080/23312041.2015.1115691.

Hao, Y., & Song, W. 2023. Research Progress on the Impact of Land Use Change on Soil Carbon Sequestration. *Land* 12(1). https://doi.org/10.3390/land12010213.

Hasan, S. S., Zhen, L., Miah, G., Ahamed, T., Samie, A. 2020. Impact of Land Use Change on Ecosystem Services: A Review. *Environmental Development* 34. https://doi.org/10.1016/j.envdev.2020.100527.

James, R. E., Echeverría, C., Oyarzún, C., Morales, L. 2018. Impact of Land-use Change on Biodiversity and Ecosystem Services in the Chilean Temperate Forests. *Landscape Ecology* 33(3). https://doi.org/10.1007/s10980-018-0612-5.

Kalnay, E., Cai, M. 2003. Impact of Urbanization and Land-use Change on Climate. *Nature* 423(6939). https://doi.org/10.1038/nature01675.

Kuswanda, W. 2019. Pengetahuan, Persepsi dan Kebijakan Pengelolaan Ekowisata Gajah di Kawasan Aek Nauli, Danau Toba. *Inovasi* 16(2). https://doi.org/10.33626/inovasi.v16i2.150.

Kuswanda, W., Sunandar, A. D. 2019. Analysis of land use change and its relation to land potential and elephant habitat at Besitang Watershed, north Sumatra, Indonesia. *Biodiversitas* (20(1). https://doi.org/10.13057/biodiv/d200141.

Ling Z., Nan, Z., Xu, Y., & Li, S. 2016. Hydrological Impacts of Land Use Change and Climate Variability in the Headwater Region of the Heihe River Basin, Northwest China. *PLoS ONE* 11(6). https://doi.org/10.1371/journal.pone.0158394.

Lyu, M., Zhou, Y., Wei, Y., Li, J., Wu, S. 2023. The Impact of Land Use Changes on Carbon Flux in the World's 100 Largest Cities. *Sustainability (Switzerland)* 15(16). https://doi.org/10.3390/su151612497.

Nugroho, P., Marsono, D., Sudira, P., Suryatmojo, H. 2013. Impact of Land-use Changes on Water Balance. *Procedia Environmental Sciences* 17. https://doi.org/10.1016/j.proenv.2013.02.036.

Pan, T., Zuo, L., Zhang, Z., Zhao, X., Sun, F., Zhu, Z., Liu, Y. 2021. Impact of Land Use Change on Water Conservation: A Case Study of Zhangjiakou in Yongding River. *Sustainability (Switzerland)* 13(1). https://doi.org/10.3390/su13010022.

Pasaribu, P. H. P., Situmorang, R. O. P., Saputra, M. H., Hartatik, S. E., Budiharta, S., Siregar, M. 2023. Analysis of Land Cover Changes Case in the Batang Toru Landscape. *IOP Conference Series: Earth and Environmental Science* 1271(1). https://doi.org/10.1088/1755-1315/1271/1/012039.

Pohan, S. A., Sulistiyono, N. 2023. Analysis of the Relationship Between Urban Heat Island (UHI) Phenomenon and Land Cover Change in Medan City Using Landsat Satellite Imagery. *Journal of Physics: Conference Series* 2421(1). https://doi.org/10.1088/1742-6596/2421/1/012017.

Qiu, H., Hu, B., Zhang, Z. 2021. Impacts of Land Use Change on Ecosystem Service Value Based on SDGs Report–Taking Guangxi as an Example. *Ecological Indicators* 133. https://doi.org/10.1016/j.ecolind.2021.108366.

Rahmawaty, Harahap, M. M., Kurniawan, H., Rauf, A. 2021a. Assessment of Land Cover Change over a Period of Five Years in Deli Serdang, North Sumatra. *IOP Conference Series: Earth and Environmental Science* 782(3), 032007. https://doi.org/10.1088/1755-1315/782/3/032007.

Rahmawaty, Harahap, M. M., Kurniawan, H., Rauf, A. 2021b. Land Use Changes Monitoring over a Period of Ten Years in Panjang Island, Pangkalan Susu, Langkat, North Sumatra. *IOP Conference Series: Earth and Environmental Scienc* 782(3), 032008. https://doi.org/10.1088/1755-1315/782/3/032008.

Rahmawaty, Harahap, M. M., Rauf, A., Kurniawan, H. 2020. Changes in Dryland Forest Cover in Several Sub-districts in Deli Serdang Regency. *IOP Conference Series: Earth and Environmental Science* 752(1), 012041. https://doi.org/10.1088/1755-1315/752/1/012041.

Rahmawaty, Rauf, A., Harahap, M. M., Kurniawan, H. 2021. Land Cover Change Impact Analysis: An Integration of Remote Sensing, GIS and DPSIR Framework to Deal with Degraded Land in Lepan Watershed, North Sumatra, Indonesia. *IOP*

Conference Series: Earth and Environmental Science 782. https://iopscience.iop.org/article/10.1088/1755-1315/782/3/032007/pdf.

Saputra, M. H., Lee H. S. 2019. Prediction of Land Use and Land Cover Changes for North Sumatra, Indonesia, Using an Artificial-Neural-Network-Based Cellular Automaton. *Sustainability (Switzerland)* 11(11). https://doi.org/10.3390/su11113024.

Schirpke, U., Tasser, T., Borsky, S., Braun, M., Eitzinger, J., Gaube, V., Getzner, M., Glatzel, S., Gschwantner, T., Kirchner, M., Leitinger, G., Mehdi-Schulz, B., Mitter, H., Scheifinger, H., ThalerDominik Thom, S., Thaler, T. 2023. Past and Future Impacts of Land-Use Changes on Ecosystem Services in Austria. *Journal of Environmental Management* 345. https://doi.org/10.1016/j.jenvman.2023.118728.

Setyo, B., Antoko, Sanudin, Sukmana, A. 2008. Perubahan Fungsi Hutan Di Kabupaten Asahan, Sumatera Utara [Forest Land Use Change (FLUC) in Asahan District, North Sumatra. *Info Hutan* 4. https://adoc.pub/balai-penelitian-kehutanan-aek-nauli-sibaganding-km-105-aek-.html.

Sipayung, M. C., Sudarsono, B., Waluddin, M. 2019. Analisis Perubahan Lahan Untuk Melihat Arah Perkembangan Wilayah Menggunakan Sistem Informasi Geografis (Studi Kasus : Kota Medan). *Jurnal Geodesi Undip* 9(1). https://doi.org/10.14710/jgundip.2020.26201.

Verawaty, Widiatmaka, I., Firmansyah, I. 2023. Modeling of Land Use and Cover Changes (LUCC) in Deli Serdang Regency, North Sumatra Province. *Jurnal Pengelolaan Sumberdaya Alam Dan Lingkungan* 13(2). https://doi.org/10.29244/jpsl.13.2.237-251.

Villarreal-Rosas, Jaramar, Jessie A. Wells, Laura J. Sonter, Hugh P. Possingham, & Jonathan R. Rhodes. 2022. The Impacts of Land Use Change on Flood Protection Services among Multiple Beneficiaries. *Science of the Total Environment* 806. https://doi.org/10.1016/j.scitotenv.2021.150577.

Wahyuni, S., Guchi, H., Hidayat, B. 2014. Analisis Perubahan Penggunaan Lahan dan Penutupan Lahan Tahun 2003 dan 2013 di Kabupaten Dairi. *Jurnal Online Agroteknologi* 2(4). https://media.neliti.com/media/publications/100734-ID-analisis-perubahan-penggunaan-lahan-dan.pdf.

Chapter 3

Forest Biodiversity, Conservation, and the Role of Livelihoods in North Sumatra

Agus Purwoko[1,2,*]
and Luthfi Aziz Mahmud Siregar[3]

[1]Department of Forestry, Faculty of Forestry, Universitas Sumatera Utara, Deli Serdang Regency, Indonesia
[2]Center of Excellence for Science and Technology of Human-Animal Coexistence, Faculty of Forestry, Universitas Sumatera Utara, Deli Serdang Regency, Indonesia
[3]Department of Agrotechnology, Faculty of Agriculture, Universitas Sumatera Utara, Medan, Indonesia

Abstract

As a province located in the tropics with almost year-round rainfall, North Sumatra has high biodiversity. Many vegetation and animal species are endemic to Sumatra Island and the North Sumatra provinces. Several umbrella species also have habitats in North Sumatra Province, with all their problems and challenges. Many local, national, and international parties have long been involved collaboratively in various efforts to conserve biodiversity and ecosystems. Many interesting cases take the form of conservation programs that can be simultaneously packaged into economic empowerment models for the surrounding community and regional economic development. In this chapter, several lessons learned are presented on how the community optimizes its social resources to be actively involved in efforts to conserve plants, animals, and their ecosystems with a management model that provides support for them, helps the economy of the surrounding community, and can even be a generator of the economic

* Corresponding Author's Email: agus9@usu.ac.id.

In: Natural Resources, Environment, and Food and Energy Security …
Editors: Mohammad Basyuni and Severino Salmo III
ISBN: 979-8-89530-283-5

growth of the related region. Lessons learned from the conservation program of species and their ecosystems in Gunung Leuser National Park (representing mountain forest ecosystems) and conservation programs and rehabilitation of mangrove ecosystems on the east coast of North Sumatra Province (representing coastal ecosystems) for ecotourism development will be explained by databases from various related agencies and research that has been carried out.

Keywords: forest, biodiversity, conservation, livelihoods, North Sumatra, economic development

Introduction

Personification is applied in the enforcement of environmental laws. Some studies on environmental law enforcement explain how the environment can become a legal subject recognized as having rights (Usman, 2018). In Indonesia, the environment is also considered a rights holder, ensuring its protection to guarantee human survival (Usman, 2018).

In the North Sumatra Province, forest biodiversity is an important element of the ecosystem that supports the lives of the surrounding community. This biodiversity not only serves as a valuable natural resource but also provides economic support and environmental sustainability (Niesenbaum, 2019). Various endemic flora and fauna live in the forests of North Sumatra, creating unique landscapes for researchers and tourists. Conservation is the key to maintaining the sustainability of biodiversity in North Sumatra. Various efforts have been made, ranging from the establishment of national parks to rehabilitation programs for damaged land. One example is the Gunung Leuser National Park, which is one of the largest conservation areas in Southeast Asia. These efforts are vital for protecting endangered species, such as orangutans and Sumatran tigers, whose populations continue to decline owing to habitat destruction, land use conflicts, and illegal poaching.

The role of biodiversity in improving the welfare of North Sumatra cannot be ignored. Many local communities depend on the natural resources in the forest, such as non-timber forest products and environmental services (Gatiso, 2019). Therefore, the government and related institutions must involve the community in sustainable forest management so that they can feel the benefits without damaging the ecosystem (Thammanu et al. 2021).

Ecotourism development is another solution for preserving biodiversity in North Sumatra while improving the local economy. By utilizing their natural potential, such as beautiful forests and rare animals, ecotourism can be an alternative source of income for surrounding communities (Kim et al. 2019). However, ecotourism development must be conducted widely, considering the principles of conservation and sustainability (Seervi, 2023).

According to Seddon et al. (2019), the biodiversity, conservation, and economic roles of communities are interrelated and must be managed holistically. By maintaining biodiversity, we also maintain the lives of the people who depend on these natural resources (Isbel et al. 2017). In North Sumatra, conservation efforts and community economic development must be conducted in a balanced manner to create harmony between humans and nature. In North Sumatra Province, forest biodiversity is not only a natural heritage that must be maintained but also has great potential to provide economic benefits to the community. For example, non-timber forest products such as fruits, spices, and traditional medicines have high selling value in both local and international markets (Masiero, Pettenella, and Cerutti, 2015). Additionally, the potential of natural tourism, such as beautiful scenery and diversity of animals, can attract tourists looking for authentic natural experiences (Winterbach et al. 2015).

According to Checina et al. (2018), conservation plays an important role in creating sustainable livelihoods for communities, particularly those who live around forests and depend on natural resources. By maintaining biodiversity and forest ecosystems, conservation helps maintain the availability of natural resources that form the basis of livelihoods, such as non-timber forest products, medicinal plants, and other forest products (Kidane and Balke, 2020). Furthermore, conservation efforts have the potential to create new jobs, such as in ecotourism, education, scientific research, and sustainable management (including utilization) of areas (Koki, 2017). Furthermore, conservation also provides opportunities for communities to develop new skills and knowledge (Brückle, 2019). Through training and educational programs related to conservation, communities can acquire new skills to increase their competitiveness in finding jobs or starting new environmentally sustainable businesses (Ardoin, Bowers, and Gaillard, 2020). Conservation is the most important way to maintain the buffering function of ecosystems for human life (Oguh et al. 2021; Pisani et al. 2021; Rawat and Agarwal, 2015). Thus, conservation is crucial not only for maintaining biodiversity but also for improving the welfare and economic development of the community as a whole (Riehl, Zerriffi, and Naidoo, 2015).

To support the sustainability of biodiversity and its benefits to communities, involving various parties, including governments, local communities, and conservation institutions, is important (Husarin et al. 2018). Establishing partnerships between these parties can strengthen conservation efforts, increase community understanding of the importance of protecting forests, and develop models of forest management that are economically, socially, and environmentally sustainable (Hvenegaard et al. 2015). Environmental education also plays a vital role in raising public awareness of biodiversity and conservation (Moss, Jensen, and Gusset, 2017). By incorporating environmental education into the school curriculum and implementing community extension programs, it is hoped that awareness of the importance of protecting the environment can increase so that the community can actively participate in efforts to preserve biodiversity in North Sumatra.

Economic Empowerment of Communities Around Forests Based on the Conservation of Species and Their Ecosystems: A Case Study in Tangkahan, Gunung Leuser National Park

The Tangkahan Nature Ecotourism Area is a tourist attraction located on one side of Gunung Leuser National Park (GLNP) in Langkat Regency, North Sumatra Province, Indonesia. Tangkahan is located in two villages: Namo Sialang and Sei Serdang. The combination of tropical rainforest vegetation and a hilly topography makes Tangkahan an ideal area for natural tourism. The Buluh and Batang Serangan Rivers that divide this forest are typical tropical forests that are still clear and flow throughout the year with low discharge fluctuations, typical of rivers that maintain an upstream catchment area. This area continues to be enriched with various types of lush and soothing tropical forest vegetation. The physical landscape in this area is decorated with beautiful cliffs and hills along its riverside (Liyushiana et al. 2023; Purwoko et al. 2021).

Biodiversity Richness and Attractiveness of Ecotourism Objects in the Tangkahan Area

The Tangkahan Ecotourism Area has many unique natural physical landscapes and biodiversity in terms of flora and fauna. The region has several spots that have natural tourist attractions (Purwoko, 2024; Rusli and Pujiwioto, 2018). A tourist attraction is a formation or related facility that attracts tourists or visitors to an area or tourist attraction (Yolanda, Yuliana, and Pramudia, 2021; Purwanto and Purwanto, 2016; Leiper, 1990). Without attractions in a place, developing tourism activities can be challenging. The attractiveness of a tour is the main capital, which is why tourists engage in tourist activities, especially natural tourism (Kurniawan, Solihin, and Putri et al. 2023). Attractiveness is a factor that makes people want to visit and see places with such attractions (Kawatu, Mandey, and Lintong, 2020; Leiper, 1990). The study of the attractiveness of a tourist attraction aims to determine the forms of recreational activities that can be developed in accordance with attractiveness and available resources.

Visitors can enjoy various ecotourism activities in Tangkahan. Ecotourism activities are the main attractions in this region and include enjoying the beauty of the natural scenery, elephant bathing, hot springs, river tubing, patrols with elephants, waterfalls, and exploring clear rivers that enter the Gunung Leuser National Park Area. The exotica of this river, which is a natural tourist attraction, is characteristic of the Tangkahan Area (Supriatna and Margules, 2022; Purwoko et al. 2021). The natural scenery that can be enjoyed in this area is green scenery from the expanse of the tropical forest in Gunung Leuser National Park. What stands out from this attraction is its beauty and coolness, as well as the beautiful landscape of the confluence of two clear rivers that flow throughout the year, namely the Batang Serangan River and the Buluh River. The Tangkahan area in the Penangga zone of Gunung Leuser National Park is rich in flora and fauna. In terms of fauna, this area is the habitat of four of the five Indonesian terrestrial umbrella species: Sumatran tigers (*Panthera tigris* sumatrae), Sumatran elephants (*Elephas maximus* sumatranus), Sumatran rhinos (*Dicerorhinus sumatrensis*), and Sumatran orangutans (*Pongo abelii*). The umbrella species have an extensive roaming area and support the success of the conservation efforts of other species. These species indirectly protected other species living within the same ecosystem. The Tangkahan area is a tropical rainforest consisting of a Dipterocarpaceae primary forest and a mixed primary forest. According to Lubis (2006), this area is dominated by plants belonging to the families Dipterocarpaceae, Meliaceae, Burceraceae, Euphorbiaceae, and Myrtaceae. This area also has a rare endemic tree in the form of the tualang tree (*Koompasia exselsa*). According to Warsodirejo (2020), tualang trees in North Sumatra are only found in Tangkahan, which is a buffer area for Gunung Leuser National Park. In addition, large trees with diameters greater than one meter, such as resin, meranti, king, and sandalwood trees, can still be seen on relatively easy paths to reach beautiful forest areas in this region.

One of the main attractions of Tangkahan ecotourism is the flow of two rivers with clear water that flows throughout the year because they are still upstream in a relatively well-maintained forest ecosystem. Tangkahan Ecotourism also has a beautiful and cool waterfall that can be used as a place for relaxation. There are 10 waterfall attractions in the Tangkahan Ecotourism Area: Sungai Buluh Waterfall, Sungai Garut, Sungai Gambir, Tala Tala, Grogoh Kiri, Alor Grogoh Kanan, Alor Simpang Kanan, Batak, Cengkeh Cengkeh, and Dan Murbe. This area also has three hot springs that are in great demand by visitors, as reported by Purwoko et al. (2021) and Robetmi and Sari (2021), namely the Buluh River Hot Spring, Sekucip Hot Spring, and Glugur River Hot Spring. According to a report by Purwoko et al. (2021), natural attractions in the Tangkahan Area as a whole are rated by visitors as being in the "interesting" category, with the highest attraction being river cruising.

Community-Based Conservation Model: Socio-Economic Transformation and Its Economic Contribution to Society

The Tangkahan Ecotourism Area, which is a forest area, has potential problems that are feared because of its increasing population and clearing forest areas for wider plantations. The establishment of the forest area as a national park in the early 1980s (Haryono et al. 2021) could not stop the unlimited wood harvesting activities between production forest areas and national parks. Previous studies have reported that for decades, timber harvesting has been a value system and local economic activity that has become a habit of residents in this region (Meijaard et al. 2012; Mangunjaya, 2006; Contreras-Hermosilla and Fay, 2006). These activities are reported to continue even when supervision is weak (Cochard, 2017; Iswanto, 2018). The revival of tourism began and was pioneered by young men and women in Namo Sialang Village and Sungai Serdang Village who wanted a change in social attitudes, economic patterns, and modernization obsessions (Rahmayani and Manurung, 2022; Rachmawati and Fountain, 2020; Suhendi and Ariantiningsih, 2013). Some of its founders stated that they were previously involved in illegal logging and forest encroachment. Thus, they realize that it is wrong to utilize the potential of natural resources for their welfare (Purwoko et al. 2021). They established the "Tangkahan Tourism Institute" as a forum to unite and organize themselves in managing ecotourism in the Tangkahan area of Gunung Leuser Leuser National Park (Wiratno et al. 2022; Siregar and Siregar, 2022; Ginting et al. 2019).

As an organization determined to carry out sustainable forest management, the Tangkahan Tourism Institute, under the leadership of Njuhang Pinem as its general chairman, collaborated by creating a Memorandum of Understanding (MOU) with the Gunung Leuser National Park Office on April 22, 2002. In the cooperation agreement, one of the critical points was the granting of management rights to the National Park, which covers an area of approximately 17,500 ha in the core zone of the GLNP to the administrative boundary of the village for ecotourism development. On the obligation side, the community is fully responsible for the security and sustainability of Gunung Leuser National Park, which borders the village areas. Furthermore, followed by the entry of NGOs (non-government organizations) in the field of environmentalists in conservation and conservation, namely the Conservation Response Unit (CRU), which joined the management of Gunung Leuser National Park sustainably and together (Purwoko et al. 2021; Suhendi and Ariantiningsih, 2013).

Ecotourism Management Based on Animal Conservation in Tangkahan and Its Contribution to Community Welfare

The existence of elephant conservation-based ecotourism activities in the Tangkahan Ecotourism Area began with the joining of the Tangkahan NGO, Conservation Response Unit (CRU), which was established to manage Sumatran elephants in the Tangkahan area. This elephant management activity was supported by the Veterinary Society for Sumatran Wildlife Conservation (VESSWIC), an NGO that plays a role in elephant services and healthcare throughout the island of Sumatra. At that time, VESSWIC handled 185 Sumatran elephants spread across six elephant training centers, including Tangkahan. In Tangkahan, eight elephants were treated and observed for all their behaviors (Purwoko et al. 2021).

According to Moraru et al. (2021), tourism has both positive and negative economic impacts. In general, these impacts are grouped according to their impact on foreign exchange receipts for the country, people's income, employment opportunities, prices and tariffs, the distribution of benefits and profits, ownership and control, development, and government revenue. For the people of Namo Sialang Village, Sei Serdang Village, and surrounding areas, elephant conservation-based ecotourism activities in the Tangkahan Area had a significant impact. The use of Sumatran elephants in Tangkahan was initially limited to the use of elephants as a means of patrolling the potential for illegal logging in the area. The elephants were cared for, fed daily, and bathed twice daily, in the morning and evening. One of the attractions of the Tangkahan Area is the moment when elephants are bathed. By performing elephant-bathing activities, visitors can interact directly with the elephants who are bathed and used as tour packages. Tourists can feel warm interactions with elephants, one of which is by feeling the water sprayed from the elephant's trunk. Bathing elephants is an important moment from the perspective of elephant in-situ conservation because, at this time, elephant handlers can establish a closer emotional connection with the elephants they care for. Finally, the activity was upgraded to a tour package, providing economic benefits to the community. In addition to officers directly involved in elephant attractions, communities whose land is passed by elephant paths during patrolling also receive a share of the distribution of economic benefits from these elephant attractions (Supriatna and Margules, 2022; Purwoko et al. 2021).

The amount of financial income also affects the absorption of a large workforce from community participation in the use of Sumatran elephants as tourist attractions. The types of jobs and businesses created include board managers, tour guides, elephant handlers (mahouts), elephant care officers, people who open businesses in the form of souvenirs, accommodation, consumption, and all related sectors. As Hunt et al. (2015) stated, community welfare is created by job opportunities and increased income through the development of ecotourism potential. This management model is a form of human-wildlife coexistence that deserves further study. The coexistence model is mutually prosperous; therefore, it becomes a win-win solution in the context of mitigating human-wildlife conflicts.

Nature Tourism Development Through Community-Based Mangrove Conservation: Lessons Learnt from the Mangrove Beach Nipah Village, Sei Nagalawan, North Sumatra

Mangrove Beach Kampung Nipah is an ecotourism area in North Sumatra, located in Sei Nagalawan Village, Perbaungan District, Serdang Bedagai Regency, and the East Coast area of North Sumatra. This is because most of the area is vegetated by mangroves. This location was initially isolated because of the inadequate and narrow road infrastructure. Transportation facilities in this area are also minimal; only private vehicles can travel, while public transportation has not yet reached the area. Communication facilities and data networks are available but are still limited/slow. Facilities in the form of signposts and public street lighting from the government remain relatively scarce (Sari et al. 2024; Purwoko et al. 2023).

This tourist attraction was pioneered from 2004–2009 independently by people in Sei Naga Lawan Village who wanted to escape poverty. The increasingly limited catch potential of the fisheries sector and the entrenched patron-client economic structure have trapped the general community in the trap of middlemen and powerful local economic groups. They are exploited economically and socially, with suffocating interest rates of up to 20–30% (Purba, Siregar, and Adela, 2019). To overcome this problem, fishermen in Sei Naga Lawan Village established a cooperative called Koperasi Serba Usaha (KSU) Muara Baimbai in 2012, which managed educational tourism and mangroves in Nipah Village. In addition, this cooperative organizes savings, loans, and the trade of marine products produced by its members (Purwoko et al. 2023). Currently, Kampung Nipah Mangrove Beach has become one of the best mangrove ecosystem-based natural attractions along the east coast of North Sumatra Province, although it is still dominated by local visitors. In addition to the attractiveness of the landscapes and mangrove ecosystems, various contemporary support facilities that follow the trend of natural tourism are available. Visitors can enjoy various natural attractions and educational tour packages. Starting in 2022, this community-based business group obtained permission from the Ministry of Environment and Forestry to be a forest area management partner for ecotourism activities. Social resources, in the form of unity, togetherness, openness, and cohesiveness, are the main forces managing this community-based tourism object (Purwoko et al. 2023; Harahap and Absah, 2022).

The Mangrove Ecosystem of Nipah Village: Characteristics, Biodiversity, and Problems

Natural conditions must be considered when managing mangrove ecotourism. Strong winds and tides affect the conditions of mangrove forests and coastlines (Koh et al. 2018). Within the last few years, the mangrove ecotourism area of Nipah Village has experienced many changes along the coastline (Harefa et al. 2020). At some point in this ecotourism area, fish and crab ponds with silvofishery patterns were built and used as attractions for visitors. Because of abrasion and tidal waves, the silvofishery area no longer receives water flowing from the sea. This is caused by the occurrence of large tidal waves that can bring beach sand to cover existing waterways and mangrove ecosystems. This condition not only affects ponds that can no longer function properly but also impacts the death of mangrove plants. Mangrove habitats that are no longer exposed to the receding seawater can dry and die. Ecotourism managers strive to reduce the impact of climate change. To protect coastal areas, ecotourism managers continue to strive to reorganize them. Mangroves that have died will be replanted in areas that continue to experience tidal flow. In addition, tidal channel engineering has been carried out to restore existing ponds and mangroves.

Purwoko et al. (2023) reported 10 species of mangrove plants in the mangrove ecosystem of Nipah Village. The growth rates of seedlings and underground plants showed the highest diversity in types and densities. This indicates good regeneration where mangrove regeneration occurs because of natural factors and human intervention (planting through educational tour packages). *Acanthus ilicifolius* is the most abundant species and an indicator of mangrove habitats that have been degraded in the past. The most abundant individual tree was Avicennia marina. This can be seen in the density of this species, which has the highest density at the level of trees and seedlings), which is as much as 248.1 and 5,238.8 ind/ha. According to the Decree of the State Minister of Environment of the Republic of Indonesia No. 201 of 2004, if the density (trees/ha) is greater than 1,500, then it is included in the category of “very high” density. This species is suitable because of the favorable growth conditions of the species (Ryck et al. 2016; Semeniuk et al. 2000). Rhizopora is widely found in this location, with a fairly high distribution, because this species is the most widely planted by managers and tourists. The ease of breeding, planting, and growth are the reasons for choosing this species for environmental education tour packages.

Garbage is a major problem in coastal areas (Chua and Garces, 1992; Saja et al. 2021). This was also reported by the mangrove ecotourism manager of Nipah Village. Although environmental hygiene programs have been conducted in tourist attraction areas through clean-behavior campaigns and the provision of local hygiene facilities, ecotourism managers still often find garbage thrown carelessly by visitors. In addition, the arrival of shipment waste carried away by currents and stuck on the beach and mangrove ecosystems in tourist attraction areas is a latent problem at this location (Lima et al. 2020; Keller and Tally, 2020). This is because littering cannot be eliminated from the community around the area, including in the upstream area of the river that empties into the nearest beach. This condition is not only an eyesore, but it can also pollute the environment. Ecotourism managers always try to appeal to visitors and not litter. The community, which is a member of the tourism object management cooperative, routinely carries out beach cleaning service work using garbage and polluting dirt.

The existence of mangroves is not only the responsibility of the tourism object manager but also of visitors and the surrounding community. The attitudes of people who generally do not care about the environment can damage the surrounding environment, especially mangrove ecosystems, through garbage pollution. Illegal logging and conversion of mangroves have occurred and continue to occur, making coastal ecosystems vulnerable to damage. Coastal areas that experience abrasion, intrusion, and deformation pose serious challenges in the development of mangrove ecosystem-based natural tourism, including Kampung Nipah Mangrove Tourism.

Improving the Community Welfare Model Through Community-Based Mangrove Tourism in Nipah Village

The presence of activists and NGOs engaged in environmental conservation is considered to have brought about a change in the mindset of the people of Sei Nagalawan Village to love the environment, especially coastal ecosystems. In this region, many cadres of environmental conservationists from local communities care for and love their environment. The presence of the "Muara Baimbai" mangrove conservation group, which later formed the "Muara Baimbai" Multipurpose Cooperation in this village, has been able to motivate the surrounding community to maintain, protect, and use the forest as a source of human life. This has persisted until now. The "Muara Baimbai" Multipurpose Cooperative is a joint business institution that resulted from the merger of two previous groups: the Baimbai Paddle Fishermen Group and the Muara Tanjung Fishermen Women's Group. Muara Baimbai was chosen because of the philosophy of the group. Muara is taken from a group of women, meaning end or end, while Baimbai means together. The word Baimbai, placed at the very back, is interpreted as a form of men's responsibility to row together to achieve one goal. Baimbai wood itself is a type of wood that can be used to make paddles. Officially, this tourist attraction opened in 2011 (Sibuea and Putra, 2021).

The manager of the mangrove forest ecotourism area in Nipah Village offers visitors mangrove education and conservation tourism (Purwoko et al. 2023). This concept allows visitors not only to enjoy the beach and the facilities available but also to provide information about mangrove forests. The manager prioritizes an open and cooperative attitude, with all parties willing to help or cooperate. Many governmental, nongovernmental, private, and academic institutions are involved in supporting the development of this tourist attraction. Various forms of support include the maintenance and addition of facilities and tourist attractions, planting and maintaining mangrove ecosystems, training and capacity building of human resources, licensing and institutional advocacy, and training and development of online promotional and marketing media. Scientific research and publications have also been conducted by various academic partners interested in studying this tourist attraction as a form of support for community-based conservation efforts.

In the early stages of development, mangrove ecotourism managers tried to promote cooperation with the local government and tourism travel agents in North Sumatra by placing advertisements and distributing brochures in strategic places. The role of media also helps to increase the number of mangrove ecotourism visitors. Mangrove ecotourism has been widely published in print media, such as national and local newspapers. Websites, such as personal and community blogs, have indirectly become a means of promotion through articles published by website owners and social media that have visited mangrove ecotourism areas. Social media platforms, such as Instagram, are considered capable of helping with promotions, with many tourists uploading photos and videos taken in ecotourism areas (Sari, Thoha, and Sadalia, 2024; Taviani, 2023).

Managerially, this mangrove ecosystem-based tourism object is managed with the principle of togetherness and mutual cooperation in accordance with the business institutional model in the form of "Cooperation," which is a distinctive business institutional model based on

Indonesian culture (Harahap and Absah, 2022). Tourist attraction service officers were selected by rotating their fellow cooperative members. If visitors are crowded (such as during holidays), all cooperative members are jointly involved in the service. Each member has their own role, ranging from ticketing officers, parking controllers, security, cleanliness, cooks, consumption waiters, accommodation, and administration to the maintenance of object facilities. Heavy work such as facility rehabilitation, embankment repairs, and mangrove planting is carried out with mutual assistance. Each working member is paid wages from the sale of entrance tickets, tour packages, consumption, accommodation, boat rentals, and other services. The remaining operating profits are collected, recorded, calculated, and reported at the annual members' meeting. The remaining proceeds recorded in each closing year are divided equally among members, usually before the Islamic holiday, which is the religion of the majority of the local population. Some of the remaining business proceeds will be agreed upon and allocated to add or improve the required facilities.

One of the service products offered at the mangrove tourist attraction in Nipah Village is a variety of seafood, whose raw materials are obtained from the catches of members of the cooperative from the fishing business unit (Purwoko et al. 2023). Thus, members' catches are sold at a higher price because they are in tourist packaging, and there are merchant margins that can be cut. If required, fishermen can obtain capital assistance for fishing activities at sea. Community members with special needs (such as parties, sending children to school, home rehabilitation, etc.) can also apply for loans with soft interest to cooperatives through savings and loan business units. Thus, cooperative members no longer have to be entangled with loan sharks and fall victim to the multivalent patron-client economic structure that develops in coastal and rural areas.

Cooperation members from housewives also utilize various raw materials that can be taken from mangrove ecosystems to be processed into various mangrove-based food products, such as *pedada* syrup (from Sonneratia caseolaris fruit), *dodol api-api* (from Avicennia marina fruit), *lindur* biscuits (from Bruguiera sp. tree fruit), and *jeruju* chips (from the leaves of Acanthus ilicifolius). Various processed mangrove food products are marketed to visitors as culinary offerings, souvenirs, and markets outside tourist attractions (Purwoko et al. 2024; 2022; 2021; Prayogo, Purwoko, and Hartini, 2015). Basically, a principle of gender equality exists in the management of mangrove tourism in Nipah Village, where all family members, both men and women, work. This food-processing activity from mangrove raw materials further strengthens the role of women in productive efforts to manage natural tourism objects while strengthening their role in mangrove ecosystem conservation.

Conclusion

Mangrove conservation efforts in North Sumatra, especially in Indonesia's high biodiversity areas such as Gunung Leuser National Park (GLNP) and mangrove ecosystems on the east coast of North Sumatra, demonstrate biodiversity conservation and social resource optimization. Programs that involve local communities can support community economic development as well as conservation efforts and provide community awareness in protecting the natural environment. In addition, mangrove rehabilitation programs can be carried out together with mangrove ecotourism development to obtain social and economic benefits.

Disclaimer

None.

References

Ardoin, N. M., Alison, W., Bowers, A., and Estelle G. 2020. Environmental Education Outcomes for Conservation: A Systematic Review. *Biological Conservation* 241:108224. doi: 10.1016/j.biocon.2019.108224.

Brückle, Irene. 2019. "The Development of Skill Knowledge in Conservation." *Restaurator. International Journal for the Preservation of Library and Archival Material* 40:287–310. doi:10.1515/res-2019-0016.

Chechin, M., Neveux, Y., Parkins, J. R., Hamann, A. 2018. "Balancing Conservation and Livelihoods: A Study of Forest-dependent Communities in the Philippines." *Conservation and Society* 16:420. doi:10.4103/CS.CS_16_182.

Chua, T., Garces, L. R. (Eds.). 1992. "*Waste Management in the Coastal Areas of the ASEAN Region: Roles of Governments, Banking Institutions, Donor Agencies, Private Sector and Communities, Vol.* 10. WorldFish.

Cochard, R. 2017. "Scaling the Costs of Natural Ecosystem Degradation and Biodiversity Losses in Aceh Province, Sumatra." *Redefining Diversity & Dynamics of Natural Resources Management in Asia,* 1(1): 231-271. doi:10.1016/B978-0-12-805454-3.00013-X.

Contreras, H., Arnoldo, Fay, C. 2006. [*Strengthening Indonesia's forest management through land tenure system reform: issues and framework for action*] World Agroforestry Centre.

Gatiso, T. T. 2019. "Households' Dependence on Community Forest and their Contribution to Participatory Forest Management: Evidence fromRrural Ethiopia." *Environment, Development and Sustainability* 21:181-197. doi:10.1007/s10668-017-0029-3.

Ginting, N., Rizky, M., Siregar, C. R., Triska, E., Ayu, P., Surya, W., Pratiwi. 2019. Study of aspects of community participation in Tangkahan ecotourism area," in *Talenta Conference Series: Energy and Engineering (EE)* 2(1). doi: 10.32734/ee.v2i1.412.

Gold, E. Compensation in Oil Pollution Damage [International Seminar Summary] in *Bandung: The ICLOS-NILOS Seminar Law of the Sea in 1990'S: Offshore Development in Indonesia.* 1991 p .261.

Harahap, R. H., Absah, Y. 2022. The study of mangrove ecotourism at Sei Nagalawan Village, Perbaungan Sub-district, Serdang Bedagai Regency, North Sumatra Province," in *2nd International Conference on Social Science, Political Science, and Humanities (ICoSPOLHUM 2021),* pp. 179-187. Atlantis Press. doi:10.2991/assehr.k.220302.026.

Harefa, M. S., Pangaribuan, B. J. T., Amri, S., Andre, K. 2020. [Analysis of mangrove forest ecosystem conservation in the coastal area of Kampung Nipah, Perbaungan District]. *Jurnal Georafflesia: Artikel Ilmiah Pendidikan Geografi* **5**:112-123. doi:10.32663/georaf.v5i2.1529.

Hartoyo, A. P., Sunkar, A., Fadillah, A., Hidayati, B., Winata, Hadi, A. N. 2021. Vegetation over analysis and ecotourism business model for sustainable forest management in Gunung Leuser National Park, Indonesia. *IOP Conference Series: Earth and Environmental Science*, 771:012001. IOP Publishing. doi:10.1088/1755-1315/771/1/012001

Hunt C. A., William, H., Durham, Driscoll, A., Martha Honey, M. 2015. Can Ecotourism Deliver Real Economic, Social, and Environmental Benefits? A study of the Osa Peninsula, Costa Rica. *Journal of Sustainable Tourism* 23:339-357. doi:10.1080/09669582.2014.965176.

Husai, M., Vishwakarma, D. K., Rathore, J. P., Rasool, A., Parrey, A. A., Mahendar, K. 2018. "Local People Strategies in Biodiversity Conservation and Sustainable Development." *The Pharma Innovation Journal* 7:444-450.

Hvenegaard, G. T., Carr, S., Clark, K., Dunn, P., Olexson, T. 2015. Promoting Sustainable Forest Management Among Stakeholders in the Prince Albert Model Forest, Canada. *Conservation and Society* 13:51-61. doi: 10.4103/0972-4923.161222.

Isbell, F., Gonzalez, A., Loreau, M., Cowles, J., Díaz, S., Hector, A., Larigauderie, A. 2017. Linking the Influence and Dependence of People on Biodiversity Across Scales. *Nature* 546:65-72. doi:10.1038/nature22899.

Iswanto, J. 2018. Kajian Hukum Lingkungan terhadap Pembangunan Jalan di Kawasan Tropical Rainforest Heritage of Sumatera (Studi Pembangunan Jalan di Karo-Langkat).

Kawatu, Victoria S., Mandey, S. L., Lintong, D. C. 2020. The Effect of Tourist Attraction on Revisit Intentions with Satisfaction as Intervening Variables at Bukit Kasih Kanonang. *Jurnal EMBA: Jurnal Riset Ekonomi, Manajemen, Bisnis Dan Akuntansi* 8. doi:10.35794/emba.v8i3.30221.

Keller, E. A., and Tally, T. 2020. Effects of Large Organic Debris on Channel Form and Fluvial Processes in the Coastal Redwood Environment. In *Adjustments of the Fluvial System*, 169-197. Routledge.

Kidane, L., Balke, A. 2020. Ethnobotanical Study on Sustainable Use of Non-Timber Forest Products and Their Source Plant Species Contribute to Balance Livelihood and Conservation: A Case of Sera Forest, Amigna District, Southern Ethiopia. doi:10.21203/rs.3.rs-21728/v1.

Kim, M., Xie, Y., Giuseppe T. Cirella. 2019. Sustainable Transformative Economy: Community-Based Ecotourism. *Sustainability* 11:4977. doi:10.3390/su11184977.

Koh, H. L., Teh, S. Y., Kh'Ng, X. Y., Raja Barizan, R. S. 2018. Mangrove Forests: Protection Against and Resilience to Coastal Disturbances. *Journal of Tropical Forest Science* 30:446-460. https://www.jstor.org/stable/26514436.

Koki, J. N. 2017. Contribution of Ecotourism Towards Sustainable Livelihood of the Communities Living on Wasini Island, Kwale County, Kenya. *Journal of Tourism and Hospitality Management* 5:106-125. doi:10.17265/2328-2169/2017.06.002.

Kurniawan, T., Solihin, R., Putri, W. A. 2023. Edu-Ecotourism Development Strategies Mapping. *Journal on Education* 5:9839-9850. doi:10.31004/joe.v5i3.1865.

Leiper, N. 1990. Tourist Attraction Systems. *Annals of Tourism Research*, 17:367-384, ISSN 0160-7383, doi:10.1016/0160-7383(90)90004-B.

Lima, A. R. A., Silva, M. D., Possatto, F. E., Ferreira, G. V., Krelling, A. P. 2020. Plastic Contamination in Brazilian Freshwater and Coastal Environments: A Source-to-Sea Transboundary Approach. in *Plastics in the Aquatic Environment-Part I: Current Status and Challenges,* pp. 427-460. Cham: Springer International Publishing. doi:10.1007/698_2020_51.

Liyushiana, Sinaga, J. C., Ismail, M., Denny, A. 2023. Assessing the Spatial Aspects of Eco-Tourism Development in Bukit Lawang Area. *International Journal of Sustainable Competitiveness on Tourism* 2:36-43. doi:10.34013/ijscot.v2i01.1241.

Mangunjaya, F. 2006. [Conservation analysis of mangrove forest ecosystem in the coastal area of Kampung Nipah Sub-district Living in harmony with nature: essays on environmental development, conservation, and biodiversity Indonesia Perbaungan]. Yayasan Pustaka Obor Indonesia.

Masiero M., Pettenella, D., Cerutti, P. O. 2015. Legality Constraints: The Emergence of a Dual Market for Tropical Timber Products? *Forests* 6:3452-3482. doi:10.3390/F6103452.

Meijaard, E., Wich, S., Ancrenaz, M., Marshall, A. J. 2012. Not by Science Alone: Why Orangutan Conservationists Must Think Outside the Box. *Annals of the New York Academy of Sciences* 1249:29-44. doi:10.1111/j.1749-6632.2011.06288.x.

Moraru AndA. D., Duhnea, C., Bărbulescu, A., Juganaru, M., Juganaru, I D. 2021. "Residents' Attitude Toward Tourism—Do the Benefits Outweigh the Downsides? The Case of Constanta, Romania." *Sustainability* 13:882. doi:10.3390/SU13020882.

Moss A., Jensen, E., Gusset, M. 2017. Impact of a Global Biodiversity Education Campaign on Zoo and Aquarium Visitors. *Frontiers in Ecology and the Environment* 15:243-247. doi:10.1002/FEE.1493.

Niesenbaum, R. A. 2019. The Integration of Conservation, Biodiversity, and Sustainability. *Sustainability* 11:4676. doi:10.3390/su11174676.

Oguh, C. E., Obiwulu, E. N. O., Umezinwa, O. J., Ameh, S. E., Ugwu, C. V., Sheshi, I. M. 2021. Ecosystem and Ecological Services; Need for Biodiversity Conservation-A Critical Review. *Asian Journal of Biology* 11:1-14. doi: 10.9734/AJOB/2021/v11i430146.

Pisani, D., Pazienza, P., Perrino, E. V., Caporale, D., Lucia, C. D. 2021. The Economic Valuation of Ecosystem Services of Biodiversity Components in Protected Areas: A Review for a Framework of Analysis for the Gargano National Park. *Sustainability* 13:11726. doi: 10.3390/su132111726.

Purba, A. M., Siregar, O. M., Adela, F. P. 2019. Digital based coastal tourism development in revolution 4.0 Era: Study on Kampung Nipah Serdang Bedagai, in *the 3rd International Conference on Social and Political Development (ICOSOP 3 2019)-Social Engineering Governance for the People, Technology and Infrastructure in Revolution Industry*, 2019, 4:264-269. doi:10.5220/0010017802640269.

Prayogo, E., Purwoko, A., Hartini, K. S. 2015. Financial Analysis of Utilization and *Processing of Jeruju Leaves (Acanthus ilicifolius L) Becomes Various Products. Thesis.* Forestry Faculty of Universitas Sumatera Utara. https://repositori.usu.ac.id/handle/123456789/54507.

Purwanto, E., Purwanto, E. 2016. *An Anti-Encroachment Strategy for* the tropical rainforest heritage of Sumatra: Toward new paradigms. Tropenbos International Indonesia, Bogor, Indonesia.

Purwoko, A., Susilawati, A., Situmorang, A. I. 2023. Assessing the Carrying Capacity of Mangroves as Raw Materials for Culinary Products: Case Study in Serdang Bedagai, North Sumatra, Indonesia. *Biodiversitas Journal of Biological Diversity* 24. doi:10.13057/biodiv/d240130.

Purwoko, A., Hartini, K. S., Rajagukguk, P. A. Financial and marketing analysis of pedada syrup (*Sonneratia caseolaris* L.) in Sei Nagalawan village, North Sumatra province, Indonesia, in *E3S Web of Conferences*, 2021, 332, p. 03006. EDP Sciences. doi:10.1051/e3sconf/202133203006.

Purwoko, A., Julianti, E., Sari, R. K., Widyastuti, D. E., Sitorus, A. K. 2024. Prospect of Processing Iindur (*Bruguiera sexangula*) Biscuits as a Culinary Product to Support the Development of Mangrove-Based Nature Tourism. *Food Research Journal.* 2024. Under Review.

Purwoko, A., Hartini, K. S., Basyuni, M., Situmorang, M. D. 2023. Community-Based Mangrove Tourism Object Development in Kampung Nipah, North Sumatra, Indonesia. *Agricultural Research* 11:241-254. doi:10.13189/ujar.2023.110203.

Purwoko, A., Thoha A. S. 2021. Analysis of the attractions of Tangkahan Nature Tourism Area, Gunung Leuser National Park, North Sumatra Province. in *IOP Conference Series: Earth and Environmental Science,* 912, p. 012065. IOP Publishing. doi:10.1088/1755-1315/912/1/012065.

Purwoko, A., Nurrochmat, D. R., Ekayani, M., Rijal, S., Garura, H. L. 2022. Examining the Economic Value of Tourism and Visitor Preferences: A Portrait of Sustainability Ecotourism in the Tangkahan Protection Area, Gunung Leuser National Park, North Sumatra, Indonesia. *Sustainability* 14:8272. doi:10.3390/su14148272.

Purwoko, A., Patana, P., Daulay, U. A. 2021. Integration of conservation and economic development in Gunung Leuser National Park in *IOP Conference Series: Earth and Environmental Science*. vol. 739. p. 012055. IOP Publishing.

Purwoko, A., Thoha, A. S., Muda. I., Sutiawan, J., Cemda, A. R., Lubis, L. H. 2024. Potential, Feasibility, and How to Develop Tangkahan Ecotourism Area, Gunung Leuser National Park, Indonesia. *Journal of Sustainable* Forestry Under Review.

Rahmayani, A., Manurung, S. S. 2022. Community Based Tourism Management at Tangkahan Tourism Attraction. *TOBA: Journal of Tourism, Hospitality and Destination* 1, no, 3:148-154. doi:10.55123/toba.v1i03.845.

Rawat, U. S., Agarwal, N. K. 2015. Biodiversity: Concept, threats and conservation. *Environment Conservation Journal* 16, no. 3:19-28. doi:10.36953/ECJ.2015.16303.

Riehl B., Zerriffi, H., Naidoo, R. 2015. Effects of Community-Based Natural Resource Management on Household Welfare in Namibia. *PLOS ONE* 10. doi:10.1371/journal.pone.0125531.

Rusli, M., Pujiwioto, T. 2018. Leuser Mountain National Park marketing strategy in optimalization of ecotourism zone," in *2nd International Conference on Tourism, Gastronomy, and Tourist Destination (ICTGTD 2018),* 2018, pp. 180-187. Atlantis Press. doi:10.2991/ictgtd-18.2018.22.

Saja, A., Aslam, A. M., Zimar, A. M. Z., Junaideen, S. M. 2021. Municipal Solid Waste Management Practices and Challenges in the Southeastern Coastal Cities of Sri Lanka. *Sustainability* 13:4556.

Sari, A., Thoha, A. S., Sadalia, I. 2024. Empowerment Strategies for the Community in the Mangrove Ecotourism Area through Creativity in Sei Nagalawan Village, Serdang. *Asian Journal of Adv.* 18(2): 31-42. doi: 10.9734/ajarr/2024/v18i2603.

Chapter 4

Local Wisdom and Sustainable Forest Management

Mahyuddin K. M. Nasution[1,*]
Siti Latifah[2]
and Opim Salim Sitompul[3]

[1]Computer Science, Fasilkom-TI, Universitas Sumatera Utara, Medan, Sumatera Utara, Indonesia
[2]Program Studi Kehutanan, Fakultas Kehutanan, Universitas Sumatera Utara, Medan, Sumatera Utara, Indonesia
[3]Data Science and Artificial Intelligence, Fasilkom-TI, Universitas Sumatera Utara, Medan, Sumatera Utara, Indonesia

Abstract

Since the beginning of time, the earth's surface has experienced shifts due to the movement of tectonic plates, the erosion of sediment, and earthquakes. Amidst these shifts, forests shift accordingly: they take root in new places to most effectively provide for surrounding organisms, becoming sustainable forests.

From an interdependence viewpoint, forests cannot be separated from human existence. Using management science, we can systematically anaylze the interactions between forests and humans. Since humans first appeared on earth, thoughts about managing forests have become part of the complex thinking required to fulfill human needs, resulting in various practices that accumulate in culture. Unfortunately, early cultures rarely recorded these thoughts; they were instead passed down through generations via stories or practices.

---

* Corresponding Author's Email: mahyuddin@usu.ac.id.

In: Natural Resources, Environment, and Food and Energy Security …
Editors: Mohammad Basyuni and Severino Salmo III
ISBN: 979-8-89530-283-5

Human communities are uniquely different from one another in their social activities. These differences in heritage and activity have been understood by researchers today as *local wisdom*. Conversely, there are written rules that relate directly or indirectly to forests and their management. The state systematically compiles these rules from human knowledge, but sometimes these written rules undermine local wisdom, which can affect how forests are protected.
To prevent this local wisdom from being lost, community members have recorded it on various, easily accessible websites. This article aims to review the types, characteristics, and features of local wisdom about forests as part of management discourse.

Keywords: science, technology, culture, rules, forbidden forest, natural law

Introduction

Forests are not only a means and source of life on Earth but have also inspired human culture (Deegen, 2024). Each forest has its own characteristics according to its geography (location on earth). Humans treat forests according to their interests or needs, or in an effort to improve human welfare and maintain a sustainable forest. This effort requires scientific management, formally known as "forest management" (Mutolib et al. 2024). The effects of this effort can vary according to the capabilities, culture, and customs of the human community living within or near the forest. In other words, local wisdom plays a key role in forest management (Sallata et al. 2024).

Written rules are created by humans to regulate behavior, and it is implemented through law enforcement. Unwritten rules, however, function to bind a collective agreement, such as natural law and cultural property, wherein compliance is implemented through ethical services or customs (Ifrani et al. 2019). While some unwritten rules can become written rules or even implemented into scientific research and theories, others remain unwritten. The unwritten rules that are passed down from generation to generation become part of the community's local wisdom (Syah et al. 2017). Depending on what is passed down to descendants and the ever-changing social environment, local wisdom can vary over time (Bengston et al. 1999). Local wisdom can also vary with the implementation of written rules. For example, if a community favoring certain methods of forest protection based on their unwritten rules is forced to implement new methods of forest management, the community's local wisdom becomes displaced by written rules (Harahap and Humaizi, 2018). The survival of local wisdom depends on how written rules are implemented as well as the community's response. Therefore, preexisting or currently existing local wisdom requires management by studying, observing, and researching criteria to reveal advantages and disadvantages. Moreover, a review is required to understand the link between forest management and local knowledge (Safa'at, 2019; Nasution, 2020a).

Forests and Local Wisdom: A Review

Culture grows and develops with human behavior. This can be seen when humans place forests as the focus of attention and everyday activity with the intention of preserving the environment. Although there are not many cultures in the world that have created and recorded scientific forest management (Muys and Messier, 2023), in every culture where forests are part of human life there is a local wisdom present that grows and develops to support life itself. For example, Indonesia, an archipelagic country consisting of 1,340 ethnic groups, has approximately 1,340 local wisdoms, each originating from a different culture (Brown and Suharsono, 1990; Nasution, 2018). Among these local wisdoms, one has the potential to have similar messages and impressions of life itself.

Inspiring Forests

Forests are a great source of life for all organisms. Forests act as water reservoirs by filtering nutrients to surrounding rivers, steams, and land (Taufik and Setiawan, 2012). In this way, the forest is a place where many lives depend. Apart from being a water reserve, forests also act as storehouses for plant seeds and fish during the dry season. Additionally, forests continuously generate air that refreshes the earth and its contents (Jena and Pradhan, 2019). Forests are home to many living species, ranging from small organisms like viruses and bacteria to large animals like the orangutan (Bruijnzeel, 1985; Morrogh et al. 2003). Different forests also contain many different trees: unique trees become the special wealth of a forest while trees that are evenly present become a balancer for the ecosystem (Srinivasa and Bhat, 2010).

Forests also inspire human communities. Forests can be separated by rivers, and islands can be separated by larger bodies of water. This creates separate regions for communities to grow. In this way, forests—due to their separate geographical locations—inspire human communities to form their own cultures (Fischer et al. 2013).

Wisdom That Records Inspiration

As a community forms, it creates its own rules that regulate interactions among members of the community and human interaction with the surrounding environment (Imtiyaaz et al. 2023). Individual habits become shared experiences and accumulate into good habits, which community members collectively consider good behavior (Lubis et al. 2023). This, over time, becomes local wisdom.

Community members are inspired by their surrounding environmental conditions such as environmental contours, land, rocks, flowing water, rivers, fish, animals, plants, foodstuffs, rice fields, fruit, vegetables, fish, and others. Each community member meets their daily needs through existing resources, which come directly or indirectly from the forest (Boedhihartono, 2022). For example, agriculture requires natural fertilizer or nutrients from forests, which naturally flow with water. Similarly, fishermen rely on fish in rivers and seas feeding on plankton, and plankton thrives due to the nutrients naturally carried by river flows from the forest (Zahroh and Sueb, 2021). Thus, both forests and resources available by mutual agreement or inheritance become subjects of local wisdom, such as forbidden forests, forbidden deep pools (Matondang, 2021), village guard livestock, and village rice fields (Mustikarini et al. 2022).

Sustainable Forests and Sustainability

The concept of sustainable forests has been around for a long time. When humans lived as nomads, they relied on forests for gathering fruit and hunting animals. As civilization advanced, humans used forests' resources to build houses, manufacture clothing, and harvest food. As such, forests have been integrated into human life for thousands of years (Radha et al. 2023).

This human-forest relationship requires protection. Thus, local wisdom is needed to manage forests sustainably (Zainal et al. 2024).

With the growth of prosperity came an encroachment of forests to expand plantations, create food estates, mines, and even build cities (Syahputra et al. 2024). Today, forests around the world, especially in Indonesia, have had their sustainability amputated not only by the government through its regulations and implementation of daily activities, but also by people who are economically and culturally poor (Vasil'ev, 1965; Rohmy and Nihayaty, 2023; Mauri et al. 2023). Forests need written rules from the government and local wisdom from the community to continue to be sustainable (Fujiwara and Onda, 2021).

Discussion

Forests have various characteristics and, depending on where and who inhabits the forest, provide different functions for the environment (Desfita et al. 2022; Bakri et al. 2023). Forests in the same regional location are divided into separate parts depending on that particular area's functionality and usefulness for both the environment and the surrounding human community. Each part of the forest provides criteria that reveal its usefulness.

Forbidden forests are areas managed by communities to protect them from human activity. These forests are preserved based on the community's local wisdom. For example, international scientific documents have recorded that the Roman *vicus* at Wareswald—a forbidden forest in Gallia Belgica—held a group of 24 bronze finger-rings attached to a wire bracelet dating back to life in the first quarter of the 4th century CE. This trinket ring was protected in the forbidden forest, for it wasn't discovered until 2018 (Kiernan and Henz, 2023).

Far from that classic place, on a different continent, in the country of Cambodia, the forest land in the Cardamom Mountains is a forbidden forest that originally produced a non-timber forest product called cardamom. This forest had been managed by the community since before the war to fertilize the lowlands as agricultural land. However, after the 1990s, the forest lost its identity: the forest that produced cardamom was encroached upon for the construction of logging roads and hydroelectric dams, agricultural expansion, and land privatization by outside parties (Hiroyuki, 2021).

Before the arrival of European nations, the archipelago Nusantara, particularly of the Minangkabau (Minang) ethnic group, had a *Rimbo Larangan* (Forbidden Forest) conservation system. The State Council created prohibitive laws to manage forests, such as not being allowed to hunt animals in the forest, cut down trees, fish in rivers, and other restrictions. However, after Indonesia declared independence in 1945, many forbidden forests in Minang land were damaged and destroyed, as they were made into production forests by the national government (Onrizal and Mansor, 2020). In contrast, the prohibited forest *Rimbo Larangan*—located in the Kenegerian Sentajo area, Riau Province, Indonesia—is managed by an institution formed by traditional stakeholders, which is more powerful than formal government institutions. The stakeholder institution manages forests according to the wishes and traditional values of the community to inspire deep awareness, a connection, and care for forest sustainability. For example, organizers conveyed several stories and myths (Berechet, 2017) and connections to Islamic values to encourage awareness, environmental responsibility and ethical practice in the surrounding community (Yasir et al. 2022).

There are many prohibited forests in Indonesia. In 2023, searching "*hutan larangan*" (forbidden forest) on Google generated over 111,000 hits (Nasution et al. 2023d), including information about prohibited forests such as the *Kampung Kuta* Prohibited Forest, Ciamis; Upper *Cibeet* River Prohibited Forest; and *Dhimbo Potai* Prohibited Forest in *Kanagarian Rumbio*, *Kampar* Regency - *Riau Province*. Some of them are forbidden forests known through oral tradition, such as the Forbidden Forest in East *Yogyakarta*, while others are recorded as documentaries presented on television, folklore, books, or filmed stories (Damşescu, 2016). This wide array of media creates a message to respect forests as part of human existence (Mills, 2008; Gupta et al. 2021).

For some humans, forests are considered sacred places for reflection, meditation, and worship. In fact, researchers have argued that the name of a forbidden forest is less often related to its vegetation, natural landscape, or geographic location than sacred reasons (Smith, 2020, Sullivan et al. 2024). Therefore, many local wisdoms are inspired by religion or spirituality. For example, the name "*hutan suci*" (sacred forest) is part of a local wisdom that emphasizes the forest as a place for rituals rather than focusing on the management of the forest with a joint-use perspective (Fowler, 2003). Therefore, the local wisdom of a forbidden forest can have a broader impact on forest management, including the surrounding environment (Shoddo, 2022).

Objectives of Sustainable Forbidden Forests

The forest is considered a source of life, as it stores water that flows to the sea through underground and surrounding rivers, carrying nutrients that fertilize agricultural land and plantations, as well as fish seeds and overflows of plankton downstream that fertilize the sea. This process provides food, supplies, and prosperity for the community, improving economic status and maintaining the environment (Henri et al. 2022).

In addition to these natural processes, the community can manage surrounding resources. For example, prohibited deep pools are implemented at several points upstream of rivers to raise fish and breed fish seeds indirectly. When the community harvests "*Lubuk Larangan*" (forbidden deep pool) every year, they indirectly benefit and bring harmony to the village (Lubis et al. 2021). To maintain food stocks and address food shortages during the transition season, village rice fields are used. Village guard animals, when used every year at communal meals, indirectly foster deliberation between community members in protecting natural resources, including forests.

From the restricted forest upstream to the community downstream, the use of shared forest resources aims to manage forests based on economic, social-community and environmental objectives. When each resource has an indistinguishable measurable estimate, it directly supports the objective of forest management based on the integration of local wisdom around forbidden forests (Kangas and Pukkala, 1992; Nasution et al. 2023a).

Constraints of Sustainable Forests

Changes in human interest in forests have created several obstacles for stakeholders interested in resources, clothing and food, raw materials for industry, oil and energy, and other luxuries related to advanced technology (Misir and Misir, 2007; Nasution, 2020c). Each stakeholder has their own criteria for forests based on their interests and personal perspectives. To avoid interference from outside interests in local wisdom, each community has a constitution that defines the identity, types and characteristics of local wisdom using multi-criteria decision making (Limaei et al. 2014; Nasution, 2020b).

Forbidden forests are protected by Article 18b paragraph (2) of the 1945 Constitution: "The state respects customary law community units and their traditional rights as long as they are still alive and in accordance with the development of society and the principles of the unitary state of the Republic of Indonesia, which are regulated in law." By showing the uniqueness of each local wisdom and using information technology and data science (Nasution et al. 2023c), it is possible to make claims that strengthen modern ways of managing forests or land that must be preserved (Firmanda et al. 2023; Nasution et al. 2023b).

Concept of Sustainable Management

When it comes to managing forests sustainbably, forbidden forests are at the core of local wisdom. As previously mentioned, other methods are forbidden deep pools, village rice fields, village guard animals, sacred forests, and others (Wadley and Colfer, 2004; Merang et al. 2020). While local wisdom follows the concepts and practices of the forbidden forest, the community, traditional institutions, and the government benefit.

Thus, to overcome the degradation of forbidden forests from external parties such as the private sector or the government, a management model that considers all the criteria and objectives of management effectively and efficiently (optimally) to support local wisdom is needed. There needs to be efforts to protect forests and life, and efforts to avoid damage and the spread of disease. Otherwise, there will be a repeat of the Lyme disease outbreak or something similar, which occurred due to forest destruction in the United States and spread to Canada (Dalmat, 2022).

Generalization of Sustainable Forest Management

Based on the management science supporting the implemnentation of the 1945 Constitution, it is possible to build a sustainable forest management model. This modeling involves determining variables based on criteria with a choice of values obtained from various measurements (Baskent and Keleş, 2009; Syah et al. 2021a). With the help of computer science, specifically statistics and artificial intelligence (Keleş & Başkent, 2011; Nasution, 2022), we can determine forest areas that are suitable for use or increase the level of sustainable forests based on local wisdom, both in terms of conservation and in producing something for human welfare (Syah et al. 2021b; Prins et al. 2023). Forests have their own criteria and objectives, formed based on benefits for humans, the environment and the forest itself. Indeed, forbidden forests are a model of forest management based on local wisdom (Zengin et al. 2015; Syah et al. 2023).

For centuries, humans have managed forests based on local wisdom. Though local wisdoms vary among geographical location, culture, and religion, they have the potential to be similar in that they support each other. (Nasution et al. 2017). Each piece of wisdom has its own characteristics that provide criteria to form a mathematical model, which aims to analyze the benefits of each policy in managing forests (Diaz-Balteiro and Romero, 2003; Nasution et al. 2020b). However, the concept of forbidden forest as local wisdom persists because every member of society benefits from its existence, either directly or indirectly. Each criterion is a variable that requires measurement and computation to determine its usefulness. Of course, all of this requires further research and study, where research and community service play a role in jointly managing sustainable forests (Turner et al. 2022; Nasution et al. 2020a).

Conclusion

Local wisdom, such as that pertaining to forbidden forests, naturally arises from the willingness of humans around the forest to manage the forest sustainably. Since the beginning of time, forests have provided resources that fulfill human needs, and, in return, humans have used local wisdom to sustainably manage the forests. With appropriate science and technology as well as local wisdom, sustainable forest management is possible.

Disclaimer

None.

References

Bakri, S., Wulandari, C., Kurniasari, N., Safei, R., Ramdani, D., Anshory. 2023. Effects of characteristics and perceptions of collectors of non-timber forest products to be members of the partnership program: A case study at Bukit Barisan Selatan National Park, Indonesia. *Jurnal Manajemen Hutan* Tropika 29(2): 109-118. DOI 10.7226/jtfm.29.2.109.

Baskent, E. Z., Keleş, S. 2009. Developing alternative forest management planning strategies incorporating timber, water and carbon values: An examination of their interactions. *Environmental Modeling and Assessment* 14(4):467-480. doi 10.1007/s10666-008-9148-4.

Bengston, D. N., Fan, D. P., Celarier, D. N. 1999. A new approach to monitoring the social environment for natural resource management and policy: The case of US national forest benefits and values. *Journal of Environmental Management* 56(3):181-193. doi 10.1006/jema.1999.0278.

Berechet, L. 2017. The midsummer night. The inhabitation of the sacred in the recuperator hermeneutics. *Revista Transilvania* 6:48-51.

Boedhihartono, A. K. 2022. Governance must evolve to meet new needs of Indonesian forest communities. *Sacred Forests of Asia: Spiritual Ecology and the Politics of Nature Conservation*, 272-291. doi 10.4324/9781003143680-25.

Brown, B. E., Suharsono. 1990. Damage and recovery of coral reefs affected by El Niño related seawater warming in the Thousand Islands, Indonesia. *Coral Reefs* 8(4):163-170. doi 10.1007/BF00265007.

Bruijnzeel, L. A. 1985. Nutrient content of litterfall in coniferous forest plantations in central Java, Indonesia. *Journal of Tropical Ecology* 1(4):353-372. doi 10.1017/S0266467400000468.

Dalmat, Y. M. 2002. Brief: Forbidden forest due to Lyme disease. *Option/Bio* 32(4): 661-662). DOI 10.1016/S0992-5945(22)00218-5.

Damşescu, A. 2016. Reflexions and reverberations of the war in the literary work of mircea eliade: The forbidden forest and twelve thousand head of cattle. *Revista de Filologia Romanica* 33101-109. doi 10.5209/RFRM.55839.

Deegen, P. 2024. The role of internal culture for coping with uncertainty in forest management. *Forest Policy and Economics* 163. doi 10.1016/j.forpol.2024.103211.

Desfita, S., Sari, W., Yusmarini, Pato, U., Budryn, G. 2022. Characteristics of forest honey from seberal areas in Riau Province, Indonesia. *Conrea Journal: Food Technology, Nutritions, and Culinary Journal* 5(2): 232-242. doi 10.20956/canrea.v5i2.912.

Diaz-Balteiro, L., Romero, C. 2003. Forest management optimization models when carbon captured is considered: A goal programming approach. *Forest Ecology and Management* 174(1-3):447-457. doi 10.1016/S0378-1127(02)00075-0.

Firmanda, H., Supaat, D. I., Wafi, M. H., Sudjito, Lestari, R. 2023. Resolving fishing disputes in lubuk larangan Kuantan Singingi, Riau Province. *Jurnal IUS Kajian Hukum dan Keadilan* 11(3):481-491. doi 10.29303/ius.v11i3.1194.

Fischer, A. P., Paveglio, T., Carroll, M., Murphy, D., Brenkert-Smith, H. 2013. Assessing social vulnerability to climate change in human communities near public forests and grasslands: A framework for resource managers and planners. *Journal of Forestry* 111(5):357-365. doi 10.5849/jof.12-091.

Fowler, C. T. 2003. The exological implications of ancestral religion and reciprocal exchange in a sacred forest in Karendi (Sumba, Indonesia). *Worldviews: Environment, Culture, Religion* 7(3):303-329. doi 10.1163/156853503322709155.

Fujiwara, T., Onda, N. 2021. Conflict of legitimacy over tropical forest lands: Lessons for collaboration from the case of industrial tree plantation in Indonesia. *Decision Science for Future Earth: Theory and Practice* 119-131. doi 10.1007/978-981-15-8632-3_5.

Gupta, K., Zhang, Y., Pai, Y. S., Billinghurst, M. 2021. VR-Wizard Towards an emotion-adaptive experience in VR. *Proceedings – SIGGRAPH Asia 2021 Posters, SA 2021.* doi 10.1145/3476124.3488657.

Harahap, H., Humaizi, H. 2018. Local wisdom in preservation of Lake Toba ecosystems (study on lake Toba community in the Village of Silalahi I, Sub district of Silahisabungan, Dairy Regency, North Sumatra Provinci). *IOP Conference Series: Earth and Environmental Science* 126(1). doi 10.1088/1755-1315/126/1/012122.

Henri, Fatansyah, F., Alita, Lestari, Y., Sonia, A., Putri, J. E., Rahmasari. 2022. Community's local wisdom and its relationship with environmental conservation efforts in Bangka Belitung, Indonesia. *IOP Conference Series: Earth and Environmental Science* 1115(1). doi 10.1088/1755-1315/1115/1/012036.

Hiroyuki, I. 2021. Land-use transformation in a highland community in western Cambodia: The historical context of clearing "forbidden forest" and rights to fallow in a cardamom production site. *Japanese Journal of Southeast Asian Studies* 59(1):146-191. doi 10.20495/tak.59.1_146.

Ifrani, Abby, F. A., Barkatullah, A. H., Nurhayati, Y., Said, M. Y. 2019. Forest management based on local culture of dayak kotabaru in the perspective of customary law for a sustainable future and prosperity of the local community. *Resources* 8(2). doi 10.3390/resources8020078.

Imtiyaaz, C. D., Wahyudiana, N. H., Nasution, E. K., Puspitasari, I. G. A. A. R., Al-Hakim, R. R. 2021. Macaque-Human interaction and the societal perceptions of Macaques in Kalisalak forest, Central Java-Indonesia. *IOP Conference Series and Environmental Science* 1220(1). doi 10.1088/1755-1315/1220/1/012022.

Jena, R., Pradhan, B. A. 2019. model to detect forest change relating to mining using Google earth engine application in Belitung Island, Indonesia. *International Conference on Space Science and Communication, IconSpace* 47-52. doi 10.1109/IconSpace.2019.8905960.

Kiernan, P., Henz, K. P. 2023. Rings from the forbidden forest: The function and meaning of Roman trinket rings. *Journal of Roman Archaeology* 36(1):73-95. doi 10.1017/S1047759423000211.

Kangas, J., Pukkala, T. A. 1992. decision theoretic approach applied to goal programming of forest management. *Silva Fennica* 26(3):169-176. doi 10.14214/sf.a15645.

Keleş, S., Başkent, E. Z. 2011. A computer-based optimization model for multiple use forest management planning: A case study from Turkey. *Scientia Forestalls/Forest Sciences* 89:87-95.

Limaei, S. M., Kouhi, M. S., Sharaji, T. R. 2014. Goal programming approach for sustainable forest management (case study in Iranian Caspian forests). *Journal of Forestry Research* 25(2):429-435. DOI 10.1007/s11676-014-0472-z.

Lubis, A. R., Nasution, M. K. M., Sitompul, O. S., Zamzami, E. M. 2023. A new approach to achieve the user's habitual opportunities on social media. *IAES International Journal of Artificial Intelligence* 12(1):41-47. doi 10.11591/ijai.v12.i1. pp41-47.Lubis, T., Dardanila, Nasution, T., Zulkarnain, Hasrul, S., Ramlan, Abus, A. F. 2021. Tradition lubuk larangan as a local wisdom for ecocultural trourism river management through landscape anthropolinguistic approach in Mandailingnese. *IOP Conference Series: Earth and Environmental Science* 926(1). doi 10.1088/1755-1315/926/1/012029.

Sallata, Markus Kudeng, Allo, Merryana Kiding, Ansari, Farji. 2024. Bamboo forest management practices and local wisdom in Tana Toraja and North Toraja Regency, South Sulawesi. *AIP Conference Proceeding* 3001(1). doi 10.1063/5.0184455.

Chapter 5

Challenges and Opportunities of Tropical Disease Management in North Sumatra

Ayodhia Pitaloka Pasaribu[1,*]
Dewi Masyithah Darlan[2]
and Fahmi Fahmi[3]

[1]Department of Pediatrics, Medical Faculty, Universitas Sumatera Utara, North Sumatra, Indonesia
[2]Department of Parasitology, Medical Faculty, Universitas Sumatera Utara, North Sumatra, Indonesia
[3]Department of Electrical Engineering, Faculty of Engineering, Universitas Sumatera Utara, North Sumatra, Indonesia

Abstract

Indonesia is the world's largest archipelagic state along the equator, making its climate almost entirely tropical. It also faces the burden of tropical diseases, such as malaria and neglected tropical diseases. North Sumatra, the fourth most populous province in Indonesia, is also dealing with the burden of tropical diseases and is currently working on eliminating them. Some diseases have a particularly high prevalence in North Sumatra. In 2023, North Sumatra ranked number four among the provinces in Indonesia in the prevalence of malaria, with some regencies still struggling to achieve elimination. Some diseases, including taeniasis, are still endemic in North Sumatra, where as other provinces eliminated them successfully years ago. Challenges in delivering optimal disease management for the affected population include a lack of infrastructure, such as point-of-care tests for

---

* Corresponding Author's Email: ayodhia@usu.ac.id.

In: Natural Resources, Environment, and Food and Energy Security …
Editors: Mohammad Basyuni and Severino Salmo III
ISBN: 979-8-89530-283-5

diagnostics, obstacles in the distribution of medical supplies and information, delayed response times resulting from ineffective disease surveillance, and difficulties in controlling outbreaks. Instability of treatment availability may make vector control strategies effective, with the added challenge of ensuring distribution to remote areas. With these challenges, there lie opportunities for medical technologists to develop innovative tools to improve management, including telemedicine and more point-of-care tests, data analytics and predictive modeling, mobile health (mHealth) solutions, public private partnerships, and wearable approaches that combine medical expertise with technological innovation.

Keywords: tropical diseases, technologies, neglected tropical disease, North Sumatra, Indonesia

Introduction

Indonesia is the third largest tropical country in the world, dominated by a tropical rainforest climate. As a result of this characteristic, the Indonesian population faces many health problems, particularly tropical diseases. (Wibawa and Satoto, 2016; Mboi et al. 2022) The Indonesian government is committed to achieving one of the targets in the Sustainable Development Goals 2030 by ending the epidemic of malaria, AIDS, tuberculosis, and neglected tropical diseases (NTDs). However, there is still a long way to go, as many diseases have not been completely eradicated, and annual outbreaks still occur (Chapman et al. 2017). Indonesia recorded nearly 400 thousand cases of malaria in 2022 (WHO, 2022). As for NTDs, taeniasis can also be found sporadically in the country (Feasey et al. 2010).

North Sumatra faces a significant public health burden due to various tropical diseases. These diseases are exacerbated by the province's diverse ecological environments and climatic conditions, which favor the proliferation of various disease vectors. Malaria remains endemic in certain parts of North Sumatra. The province's dense forests and numerous water bodies provide ideal breeding grounds for mosquitoes, contributing to persistent malaria cases and occasional outbreaks (Wibawa and Satoto, 2016). Taeniasis, an NTD, is an infection caused by tapeworms of the genus *Taenia.* This condition is most commonly caused by consuming undercooked or contaminated pork and beef in which the larvae have not been adequately killed. Taeniasis has also been reported in North Sumatra.

The province's tropical climate supports the survival and proliferation of tapeworm eggs in the environment, exacerbating the risk of infection (Wandra et al. 2020).

Challenges

North Sumatra's battle against tropical diseases such as malaria, dengue, soil-transmitted helminth, and taeniasis is made more difficult by unique challenges, including the availability and implementation of technology, the diagnostic skills of healthcare professionals, the distribution and efficacy of medicines, and other logistical and infrastructural issues. Addressing these multifaceted challenges requires a comprehensive understanding and strategic planning to effectively manage and mitigate the impact of these diseases (Ackley et al. 2021).

Accessibility and Accuracy of Diagnostic Tools

The geographical limitations (terrain and remoteness) and infrastructural limitations (road quality, health facilities, electricity, internet. and trained personnel) significantly affect the accessibility of modern diagnostic tools in remote areas of North Sumatra. Furthermore, the available tools may not always be sensitive or specific enough to detect certain tropical diseases reliably, leading to misdiagnosis and inappropriate treatment. Developing and distributing robust, easy-to-use, and low-cost diagnostic kits specifically designed for tropical environments and investment in mobile diagnostic laboratories could help bridge the gap in hard-to-reach areas (Bharadwaj et al. 2021).

Supply Chain Inefficiencies in Medicine

The distribution of medicines in North Sumatra faces considerable logistical challenges due to poor transportation infrastructure and erratic supply chains, particularly in the more isolated regions such as Nias, Pakpak barat, and Toba Samosir. Delays in patients getting essential medications can be detrimental during disease outbreaks. This can be solved by strengthening the local supply chain through improved logistic networks and local stockpiling of essential medicines (Souza et al. 2020).

Lack of Robust Health Information Systems

Inadequate health information systems in North Sumatra (fragmented data resources, paper-based records, insufficient digital infrastructure, lack of training for health workers, delayed reporting, inconsistent data entry, and underreporting of data) impede effective data collection and analysis, which is essential for managing disease outbreaks and formulating public health strategies. The solution could be to develop integrated health information systems to efficiently collect, store, and analyze health data. Partnering with tech companies to tailor these systems to the region's specific needs can enhance their effectiveness (Epizitone et al. 2023).

Opportunities

The management of tropical diseases in North Sumatra through technology faces numerous challenges that span diagnostic, medical, and infrastructural aspects. However, it also presents many opportunities that can be realized through a collaborative approach involving multidisciplinary experts. Advancements in health technology, such as digital surveillance systems, telemedicine, and mobile health applications, present unprecedented opportunities to enhance disease monitoring, improve diagnostic accuracy, and expand access to medical care. Dealing with these diseases effectively requires a synergistic approach integrating technology with robust public health strategies. This approach not only promises to reduce the disease burden but also to enhance overall public health resilience in the region (George et al. 2019).

Disease Surveillance

Advanced technology can play a significant role in disease surveillance by improving data collection, analysis, and communication, especially for tropical diseases. Using remote sensing technologies such as satellite imagery combined with geographic information systems (GIS) can help track environmental factors like temperature, humidity, rainfall, and vegetation cover. Changes in these factors can indicate potential outbreaks of diseases like malaria and dengue. GIS also helps map disease spread and identify high-risk areas (Fahmi et al. 2022; Pasaribu et al. 2021).

As in big data analytics, advanced data analytics techniques can process large volumes of data from various sources, including electronic health records, social media, and internet searches, to detect patterns and trends. For example, analyzing Google search trends or social media posts can provide early indicators of disease outbreaks in specific regions.

Another possible tool for disease surveillance is mobile health (mHealth) applications, which can facilitate real-time data collection, reporting, and monitoring of disease cases. Health workers in remote areas can use these apps to report suspected cases, track outbreaks, and receive guidance on diagnosis and treatment (Bates and Zumla, 2015). Telemedicine may also allow healthcare professionals to remotely diagnose and treat patients, which is particularly beneficial in areas where access to healthcare services is limited. It can facilitate early detection and management of tropical diseases, reducing the burden on healthcare systems.

The development of artificial intelligence and machine learning can also contribute to disease surveillance by analyzing complex data sets to identify patterns and predict disease outbreaks. They can also assist in identifying potential vectors and reservoirs of diseases, aiding in targeted control and prevention measures.

Technological advancements have led to the development of rapid diagnostic tests for various tropical diseases, allowing for quick and accurate diagnosis even in resource-limited settings. These tests enable timely intervention and treatment, reducing the spread of diseases (Pasaribu et al. 2022). By leveraging these advanced technologies, we can improve disease surveillance efforts, enhance early detection of outbreaks, and implement targeted interventions to effectively control the spread of tropical diseases.

Vector Control

There are some common vectors in North Sumatra, including mosquitoes (Aedes aegypti, Anopheles, Culex), Ixodid ticks, Xenopsylla cheopis, and trombiculid mites. There are some common vectors in North Sumatra, including mosquitoes (Aedes aegyptu, Anopheles, Culex), Ixodid ticks,

Xenopsylla cheopis and trombiculid mites. Some of these promising advanced technologies can be used for vector control in tropical disease management, as well as in monitoring, tracking, and managing disease-carrying vectors. For example, drones and unmanned aerial vehicles equipped with cameras and sensors can efficiently survey large areas to locate breeding sites, monitor vector populations, and assess the effectiveness of control measures. Drones can access remote or inaccessible areas, making vector control efforts more comprehensive.

Advanced biotechnological methods, such as genetic modification or gene editing techniques like CRISPR, can be used to develop sterile or genetically modified vectors that reduce or suppress wild populations. This approach can help control vector-borne diseases without the use of chemical insecticides.

The following are some examples of advanced biotechnological methods that have been used in North Sumatra, Indonesia:

1. Tissue Culture for Agricultural Improvement: Advanced tissue culture techniques have been used in North Sumatra to improve the quality and yield of important crops such as oil palm and rubber. This method allows for the rapid propagation of plants with desirable traits, ensuring uniformity and resistance to diseases.
2. Biofertilizers and Biopesticides: Biotechnology has been employed in the development of biofertilizers and biopesticides in North Sumatra to promote sustainable agriculture. These biotechnological products help enhance soil fertility and control pests without the adverse environmental impacts associated with chemical fertilizers and pesticides.
3. Genetic Improvement of Livestock: Advanced genetic techniques have been used to improve livestock breeds in North Sumatra. These methods include artificial insemination and selective breeding to enhance traits such as milk production in cattle, growth rates in poultry, and disease resistance.
4. Aquaculture Enhancement: Biotechnology has also been employed in the aquaculture sector in North Sumatra, particularly in the breeding and management of fish species like tilapia and catfish. Techniques such as genetic selection and disease diagnostics have improved fish yield and health.
5. Biogas Production: In rural areas of North Sumatra, biogas technology has been implemented to convert agricultural waste and

animal manure into biogas. This renewable energy source provides a sustainable solution for energy needs and helps in waste management.

6. Disease Diagnostics and Vaccine Development: Advanced biotechnological methods have been used in the healthcare sector for the diagnosis of diseases and the development of vaccines. Techniques such as polymerase chain reaction (PCR) and enzyme-linked immunosorbent assay (ELISA) are employed for rapid and accurate disease detection.
7. Plant Genetic Conservation: Biotechnology has been used in the conservation of North Sumatra's rich biodiversity. Techniques like DNA barcoding and cryopreservation help preserve the genetic material of endangered plant species for future research and restoration efforts.
8. Fermentation Technology: This technology is widely used in North Sumatra for the production of traditional fermented foods and beverages. Biotechnology has enhanced the fermentation processes, improving the quality and safety of products like tempeh, tofu, and traditional alcoholic beverages.

These examples illustrate how advanced biotechnological methods have been applied in various sectors in North Sumatra to promote agricultural productivity, environmental sustainability, and public health.

The development of smart traps equipped with sensors and wireless communication capabilities can monitor vector populations in real time (Zulkuply et al. 2018). These traps can detect changes in population density and provide early warning alerts to public health authorities, allowing for timely intervention. By integrating these advanced technologies into vector control strategies, more effective and sustainable methods for controlling the spread of tropical diseases carried by vectors like mosquitoes, ticks, and flies would be applicable.

Outbreak Control

One aspect of tropical disease management is outbreak control, and advanced technology can help by facilitating rapid response, efficient communication, and targeted intervention. Real-time data analysis tools can process large volumes of data from various sources such as hospitals, clinics, laboratories, and public health agencies. Analyzing this data in real time can help identify patterns and trends indicative of an outbreak, enabling swift response measures, including the development of predictive modeling techniques, where machine learning algorithms can analyze historical data to predict the spread of tropical diseases and identify high-risk areas.[17] Public health authorities can allocate resources more effectively by forecasting outbreak trajectories and proactively implementing preventive measures.

Mobile applications and social media (Instagram, Twitter/X, WhatsApp, etc.) alerts can be used to disseminate information about outbreak alerts, prevention strategies, and treatment options to affected communities. These tools also enable two-way communication, allowing individuals to quickly report symptoms or seek medical assistance. Telemedicine platforms enable healthcare professionals to remotely diagnose and treat patients, reducing the risk of disease transmission in healthcare settings. In addition, teleconsultation services can provide medical advice to individuals in remote areas with limited access to healthcare facilities.

GIS and remote sensing technologies can map the spread of diseases in real time, identify transmission hotspots, and assist in targeted vector control efforts. GIS can also help in resource allocation by visualizing areas with the highest disease burden (Sipe et al. 2003). By harnessing these advanced technologies, we can expect to maximize outbreak control efforts, enabling faster response times and more targeted interventions and, ultimately, mitigate the impact of tropical disease outbreaks on public health.

Conclusion

Tropical diseases are still a burden in Indonesia, particularly in North Sumatra. Despite several efforts, cases still occur, and the prevalence of some diseases, such as malaria, has rather increased. Controlling these diseases and achieving elimination requires the involvement of multidisciplinary experts, not only from medical but also from technological backgrounds. While there are numerous challenges to overcome, there are also many opportunities for better tropical disease management in North Sumatra.

Disclaimer

None.

References

Ackley, C., Elsheikh, M., Zaman, S. 2021. Scoping review of neglected tropical disease interventions and health promotion: A framework for successful NTD interventions as evidenced by the literature. *PLoS Neglected Tropical Diseases* 15(7): e0009278. doi: 10.1371/journal.pntd.0009278.

Bates, M., Zumla, A. 2015. Rapid infectious diseases diagnostics using Smartphones. *Annals of Translational Medicine 3*(15): 215. doi: 10.3978/j.issn.2305-5839.2015.07.07.

Bharadwaj, M., Bengtson, M., Golverdingen, M., Waling, L., & Dekker, C. 2021. Diagnosing point-of-care diagnostics for neglected tropical diseases. *PLoS Neglected Tropical Diseases* 15(6): e0009405. doi: 10.1371/journal.pntd.0009405.

Chapman, P., Siri, J., Chisholm, E., Chapman, R., Doll, C. N., Capon, A. 2017. SDG 3: Ensure healthy lives and promote well-being for all at all ages. *A Guide to SDG Interactions: From Science to Implementation. Paris, France: International Council for Science*, 81-126.

Epizitone, A., Moyane, S. P., Agbehadji, I. E. 2023. A systematic literature review of health information systems for healthcare. *Healthcare* (Basel, Switzerland) 11(7): 959. doi: 10.3390/healthcare11070959.

Elyazar, I. R., Hay, S. I., Baird, J. K. 2011. Malaria distribution, prevalence, drug resistance and control in Indonesia. *Advances in Parasitology* 74: 41-175. doi: 10.1016/B978-0-12-385897-9.00002-1.

Fahmi, F., Pasaribu, A. P., Theodora, M., Wangdi, K. 2022. Spatial analysis to evaluate risk of malaria in Northern Sumatera, Indonesia. *Malaria Journal* 21(1): 241. doi: 10.1186/s12936-022-04262-y.

Feasey, N., Wansbrough-Jones, M., Mabey, D. C., Solomon, A. W. 2010. Neglected tropical diseases. *British Medical Bulletin* 93(1): 179-200. doi: https://doi.org/10.1093/bmb/ldp046.

George, D. B., Taylor, W., Shaman, J., Rivers, C., Paul, B., O'Toole, T., Johansson, M. A., Hirschman, L., Biggerstaff, M., Asher, J., Reich, N. G. 2019. Technology to advance infectious disease forecasting for outbreak management. *Nature Communications* 10(1): 3932. doi: 10.1038/s41467-019-11901-7.

Mboi, N., Syailendrawati, R., Ostroff, S. M., Elyazar, I. R., Glenn, S. D., Rachmawati, T., Mokdad, A. H. 2022. The state of health in Indonesia's provinces, 1990–2019: A systematic analysis for the Global Burden of Disease Study 2019. *The Lancet Global Health* 10(11): e1632-e1645.

Pasaribu, A. P., Tsheten, T., Yamin, M., Maryani, Y., Fahmi, F., Clements, A. C. A., Gray, D. J., Wangdi, K. 2021. Spatio-temporal patterns of dengue incidence in

Medan City, North Sumatra, Indonesia. *Tropical Medicine and Infectious Disease* 6(1): 30. doi: 10.3390/tropicalmed6010030.

Pasaribu, A. P., Nasution, I. S., Sembiring, K., Fahmi, F., Pasaribu, S. 2022. Comparison of the performance of the CareStart Malaria Pf/Pan Combo test and field microscopy in the diagnosis of Plasmodium vivax malaria in North Sumatra, Indonesia. *Malaria Journal* 21(1): 27. doi: 10.1186/s12936-022-04057-1.

Sipe, N. G., Dale, P. 2003. Challenges in using geographic information systems (GIS) to understand and control malaria in Indonesia. *Malaria Journal* 2(1): 36. doi: 10.1186/1475-2875-2-36.

Souza, A. A., Holloway, C., Williams, T. 2020. The NTD Supply Chain Forum—strengthening the backbone of NTD programs. *PLOS Neglected Tropical Diseases* 14(11): e0008818. doi: 10.1371/journal.pntd.0008818.

Wibawa, T., Satoto, T. B. 2016. Magnitude of neglected tropical diseases in Indonesia at postmillennium development goals era. *Journal of Tropical Medicine 2016,* 5716785. doi: 10.1155/2016/5716785.

Wandra, T., Darlan, D. M., Yulfi, H., Purba, I. E., Sato, M. O., Budke, C. M., Ito, A. 2020. Soil-transmitted helminth infections and taeniasis on Samosir Island, Indonesia. *Acta Tropica, 202*, 105250. doi: 10.1016/j.actatropica.2019.105250.

World Health Organization. 2022. *World Malaria Report 2022.* World Health Organization: Geneva.

Zulkiply, N., Audah, L., Wahab, M. H. A. 2018. Smart Mosquitoes Trap Using Favoriot Monitoring System. *2018 IEEE Conference on Wireless Sensors (ICWiSe),* Langkawi, Malaysia, 2018, pp. 83-88.doi: 10.1109/ICWISE.2018.8633282.

Chapter 6

Contributions of Watershed Management to Land Sliding, Erosion and Flooding in North Sumatra

Abdul Rauf[1,*]
and Ramadhan Hamdani Harahap[2]
[1]Department of Agrotechnology, Faculty of Agriculture, Universitas Sumatera Utara, Medan, Indonesia
[2]Faculty of Social and Political Science, Universitas Sumatera Utara, Medan, Indonesia

Abstract

A watershed is a unitary land area with a river ecosystem and its tributaries. It collects, stores and channels rainwater to a lake or sea naturally, forming a land boundary or topographical divider and a sea boundary that marks the water area. Watersheds are often affected by land activities and simulatanously cause land activities, so they must be managed carefully and integratedly to prevent land degradation in the form of erosion, landslides, floods and/or drought, especially in wet tropical areas like North Sumatra, Indonesia. North Sumatra Province, with a total land area of 72,981.23 km^2, is divided into 228 watersheds, with very diverse characters influenced by the Bukit Barisan mountainous region, and spread from the east coast to the west coast of Sumatra, including Nias Island. There are three main factors causing watershed degradation, namely: (1) Natural conditions (geomorphology) that are vulnerable to degradation; (2) Climate, especially high rainfall; and (3) Human activities that are uncontrolled and exceed the capacity or carrying capacity of the land. In order to

* Corresponding Author's Email: a.rauf1@usu.ac.id.

In: Natural Resources, Environment, and Food and Energy Security …
Editors: Mohammad Basyuni and Severino Salmo III
ISBN: 979-8-89530-283-5

avoid damage to watersheds in North Sumatra, integrated management is needed involving the parties and the application of land use technology based on watershed management in all areas, both in forest areas, cultivation areas and residential areas.

Keywords: watershed, degradation, controlling, area type

Introduction

The North Sumatra Province region has a total land area of 72,981.23 km2 divided into 228 River Watersheds (DAS), 183 of which are in the BPDAS Asahan Barumun working area (eastern part of North Sumatra up to the border with the Province Riau and West Sumatra), and the remainder (45 watersheds) are in the BPDAS Wampu Sei Ular working area (western North Sumatra to the border with Aceh Province) (BPDAS Asahan Barumun, 2009b).

The biophysical characteristics of almost all watersheds in North Sumatra are determined by the altitude of the region starting from the East Coast to the West Coast of Sumatra, including the peaks of the Bukit Barisan mountains, both those leading to the Sumatra Strait and those leading to the Indian Ocean. This regional coverage with very diverse characteristics makes it very possible for forest and land degradation to occur, coupled with geographical conditions that are vulnerable to damage, extreme wet climatic conditions (high rainfall), and human factors that generally lack or do not understand management technology and land use based on watershed management (BPDAS Asahan Barumun, 2009a).

Watershed management is very important in all aspects of land use in North Sumatra because the entire area is in the wet tropics with average rainfall ranging from 1,431 mm to 2,265 mm per year and an average number of rainy days of 173-230 days per year. This high rainfall is the cause of all controlling processes (addition, subtraction, transfer and change of materials and energy) on the surface of the earth caused by water. A disturbed balance between controlling processes will cause land and environmental degradation, such as floods, drought, erosion and sedimentation and even landslides (BPDAS Asahan Barumun, 2003).

Therefore, watershed management, which has the main aim of accommodating, storing and channeling water from rainfall to lakes or the sea naturally, is a priority in order to control land degradation. Efforts that must be made are to regulate the reciprocal relationship between natural resources and humans in each land area and regulate all human activities, so that ecosystem sustainability and harmony can be achieved and the benefits of natural resources for humans themselves are increased in a sustainable manner. In this way, watershed management can still obtain environmental benefits in the form of sustainability and harmony of the watershed ecosystem, by carrying out the function of collecting, storing and channeling rainwater in a balanced manner, in addition to getting socio-economic benefits in the form of goods and services, avoiding floods, drought and famine, disease outbreaks, etc. on an ongoing basis (Arsyad, Pryanto dan Nasoetion, 1985; Asdak, 2002).

Factors Causing Watershed Degradation

There are at least three main factors causing watershed degradation, namely: (1) Natural conditions (geomorphology); (2) Climate; and (3) Human activities. Natural conditions (geomorphology) in the form of geology, soil and topography which are susceptible to erosion, floods, landslides and drought are one of the main causes of watershed damage. This is related to the land capacity/carrying capacity of the region and the environment against watershed damage. Climate, especially high rainfall, has the potential to cause damage to land through the role of high erosivity. Apart from erosion, sedimentation also occurs in water bodies, and on rocky sloping land it can cause landslides (land slides) (Arsyad, 2006).

Human activities in the utilization/use of land/forests are an important causal factor in land and environmental (DAS) degradation. Human activities that exceed the carrying capacity of the region/environment and/or do not apply soil and water conservation principles (due to lack of knowledge and skills) are the main causes of watershed damage. Damage to watersheds is increasingly exacerbated by the mental attitude (moral hazard) of humans in exploiting land resources uncontrollably and excessively, without paying attention to the function of the area, capabilities and suitability of the land. For this, effective and continuous efforts are needed to transfer knowledge and technology, especially land use technology based on watershed management (Asdak, 2002; Arsyad, 2006).

Some human activities that can exceed the carrying capacity of the land and become uncontrolled include (Asdak, 2002; Arsyad, 2006): (1) Deforestation: Extensive cutting down of forests to produce wood, expand agricultural land, or for other industrial purposes. This can cause ecosystem damage, loss of flora and fauna habitat, and soil erosion; (2) Uncontrolled urbanization: The increase in human population in large cities often leads to uncontrolled development. Clearing land for settlements, infrastructure and industry has an impact on land destruction, loss of agricultural land and separation of natural habitats; (3) Overgrazing: Raising livestock with an intensity that exceeds the carrying capacity of the land. These actions can cause land degradation, loss of natural vegetation, and erosion (4) Excessive use of water: Uncontrolled use of water, especially for agricultural irrigation and industrial purposes, can cause a decrease in groundwater levels and a decrease in water quality. This can threaten the survival of water resources and the ecosystems that depend on them; and (5) Excessive use of chemicals: The use of pesticides, chemical fertilizers, and other chemicals in agriculture or industry, if uncontrolled, can pollute soil and water. In the long term, this can damage the health and quality of the land.

Principles of Watershed Management

Watershed management is defined as human efforts to regulate the reciprocal relationship between natural resources and humans in the watershed and all its activities, in order to realize sustainability and harmony of the ecosystem and increase the benefits of natural resources for humans in a sustainable manner (PP No. 37/2012). Thus, the objectives of watershed management include at least 4 (four) things, namely: (1) Creating optimal water management conditions in the watershed including quantity, quality and distribution in space and time; (2) The realization of productive land conditions in accordance with the carrying capacity and capacity of the watershed; (3) Realization of better awareness, ability and active participation from parties in watershed management; and (4) Realizing improvements in community welfare.

The most important part of watershed management is controlling human activity (humans as objects) so that it does not exceed the carrying capacity of the land. Conservation efforts, use of sustainable technologies, and policies that promote sustainable agriculture and ecologically wise development can help minimize these negative impacts. On the other hand, humans as subjects are very important in watershed management, through several methods, including: (1) Raising environmental awareness: Communities must increase awareness about the importance of watersheds and the positive impact of good management. This can be done through outreach and education campaigns in schools, mass media, and community activities; (2) Environmental quality control, especially in waste management: The community must be involved in safe waste disposal and reducing plastic waste. The use of cloth shopping bags, recycling kits, and reducing the use of single-use plastic materials are some of the steps that can be taken; (3) Reforestation: Communities must be involved in tree planting and vegetation maintenance programs around the watershed. This helps in preventing soil erosion and maintaining water quality; (4) Wise use of water: People must use water wisely and not use it excessively or wastefully. In terms of implementing irrigation practices, they must also be efficient and prioritize essential water needs; (5) Collaboration: Communities should work together with local governments and NGOs to develop sustainable watershed management plans. Active involvement in community meetings and discussion forums can ensure that community voices are heard and taken into account: (6) Monitoring and reporting: Communities can carry out environmental monitoring to ensure that illegal or detrimental activities related to watersheds are reported to the competent authorities. They can also participate in research projects and present their findings. By actively participating in watershed management, communities can help preserve important natural resources and improve their own quality of life (Asdak, 2002; Arsyad, 2006).

In order to achieve the above objectives, the basic watershed management policy must be (Asdak, 2002): (1) Holistic, integrated, planned and sustainable; (2) The target area must be in its entirety from upstream to downstream areas; (3) Implementation must be decentralized and responsible while still referring to the watershed approach as a management area unit; (4) Involves participation and is based on consultation with the community at each stage of management; and (5) In extracting operational funds, management can involve the business world through a beneficiary pays principle scheme; and or polluter pays principle; cooperation funds between institutions and/or central and regional governments through cost sharing schemes; as well as voluntary funds (voluntary pays principle) from parties including from the community (individuals and/or groups).

Watershed management must be integrated, because (Asdak, 2002): (1) Watersheds as a single hydrological ecosystem have links between upstream, midstream and downstream activities; (2) Watershed components in the form of natural resources that are interrelated and influence each other; (3) Within the watershed there are many parties/sectors and communities with various types of interests; (4) Generally, watershed areas cross government administrations, including districts, provinces, and some even cross countries; (5) All scientific disciplines/professions have an interest in managing and improving watershed conditions; (6) Every person who is also a component of a watershed is responsible for maintaining sustainability and repairing damage to the watershed.

Therefore, the principles of integrated watershed management (Asdak, 2002): (1) Demand "one plan and one management system" because the watershed is a single ecosystem; (2) Involving stakeholders, coordinated, comprehensive and sustainable; (3) Adaptive to dynamic changes in conditions according to the characteristics of the watershed; (4) Implemented by dividing tasks and functions, costs and benefits between stakeholders fairly; and (5) Based on the principle of accountability.

Apart from that, integrated watershed management (DAS) is an approach that integrates various aspects of watershed system management. This approach includes monitoring and assessing watersheds, planning, controlling and monitoring and evaluating watershed management activities and so on. Integrated watershed management also aims to achieve a balance between the interests of water users and water resources in the watershed while still paying attention to environmental sustainability and conservation. The main component in integrated watershed management is water management which includes regulating water flows, controlling floods, maintaining and cleaning water channels, and providing sufficient water supplies for human, agricultural, industrial and environmental needs. Water management in relation to watersheds is the process of planning, monitoring and managing water resources in an area related to the river system and water flows within it. The main objective of water management in relation to watersheds is to maintain the balance of the water ecosystem, minimize negative impacts on the environment, and meet the water needs of the community (Asdak, 2002).

Several aspects related to water management in relation to watersheds include (Dunne, 1978; Asdak, 2002): (1) Determination of water needs: Water management planning in watersheds involves collecting data and analysis regarding water needs to ensure that the available water supply is sufficient to meet the needs of society, agriculture, industry and environment; (2) Protection of water resources: Watershed management also involves the protection of water resources. This includes controlling pollution, preventing soil erosion, and restoring damaged forests to maintain good water quality and availability; (3) Water flow regulation: Water flow regulation activities in the watershed are carried out to overcome the risk of flooding, reduce erosion and sedimentation, and maintain water quality. This can involve dam construction, construction of irrigation canals, and regulation of river flows; (4) Community participation: It is important to involve the community in water management in relation to watersheds. Through active community participation, water management policies and actions can be more targeted, transparent, and respond directly to community needs.

Water management in relation to watersheds is a complex effort and requires cooperation from various parties, including government, society and the private sector. By maintaining sustainable water management in the watershed, we can ensure the availability of sufficient water for human life and the surrounding environment. A river basin (DAS) is an area that plays an important role in the hydrological cycle and human life. Therefore, land conservation in relation to watersheds is very important to ensure survival and environmental quality (Dunne, 1978; Asdak, 2002; Arsyad, 2006).

In line with this, the scope and activities of integrated watershed management include (Asdak, 2002): (1) Land use planning to meet various needs for goods and services as well as environmental sustainability; (2) Implementation of water resource conservation to reduce the destructive power of water and to produce water (water yield) through optimizing land use; (3) Land and vegetation management inside and outside forest areas (utilization, rehabilitation, restoration, reclamation and conservation); (4) Development and management of artificial resources, especially those related to soil and water conservation; and (5) Community empowerment and development of watershed management institutions, including through the DAS Forum.

Table 1. Mechanical techniques for land use based on watershed management that have been implemented in North Sumatra

Technology	Description	Implementation Function	Watershed Management Function	Regional Suitability
Cirkle partspaying	A depression 10-15 cm deep, clean of weeds, 3-4 meters in diameter encircles the base of the oil palm tree trunk. Estimated static volume: 125-170 m^3/Ha.	Placement of fertilizer and accommodating bunches and harvested beans	Collects, stores and absorbs rainwater	Cultivation area
Big hole	A large planting hole measuring at least 1x1x1 meter is filled with compost before planting the seeds, especially in second generation oil palm cultivation and beyond. Estimated static volume: 125-135 m^3/Ha.	Prevention of Ganoderma development and microclimate adaptation	Collects, stores and absorbs rainwater	Cultivation and forest areas (Silviculture)
Hole in hole	Planting hole 40 x 40 x 30 cm in a large hole measuring at least 100 x 100 x 30 cm, filled with compost after the tree seed (usually oil palm) is planted in the small hole. Estimated static volume: 100-125 m^3/Ha.	Pencegahan perkembangan Ganoderma dan meningkatkan kelembaban di sekitar perkaran bibit	Collects, stores and absorbs rainwater	Cultivation and forest areas (Silviculture)
Rorak	The ditch is 0.5-0.6 meters wide, 0.5-0.7 meters deep, the length depends on the length of the land in the direction of cutting the slope (parallel to the contour). In oil palm plantations, it is made downstream of the harvesting area. Estimated static volume: 25-100 m^3/Ha.	Holds and diverts runoff water when it rains.	Holds, stores and absorbs water when it rains.	Cultivation and forest areas (Silviculture)
Vertical Mulch	The immersion of organic material from plant residues and/or manure into trenches or rakes that are deliberately made between rows of plants along a path cutting across the slope. Rorak/ditch also measures 0.5-0.6 meters, 0.5-0.7 meters deep, length depending on the length of the land in the direction of cutting the slope. Estimated static volume: 100-200 m^3/Ha.	Ensuring plant residues (harvest residue or pruning residue) so that they do not scatter on the ground where they can become hosts for pests and plant diseases.	Increase infiltration capacity, store and absorb rainwater.	Cultivation and forest areas (Silviculture)

Technology	Description	Implementation Function	Watershed Management Function	Regional Suitability
Rainwater Storage Tank	The ground hole is watertight (the walls are concreted or covered with plastic), the minimum hole size is 2x2x2 meters, in the shape of a well. Estimated static volume: 8 m^3/unit.	Collecting rainwater on land far from water sources for the purpose of dissolving fertilizers and pesticides	Harvesting rainwater	Cultivation and Settlement/urban areas
Infiltration wells	A hole in the ground measuring at least 2x2x2 meters in the form of a well that can carry surface runoff water into the ground. Estimated static volume: 8 m^3/unit.	Reduces puddles when it rains	Passing rainwater to the aquifer (underground water)	Cultivation, Settlement/urban and Forest Areas
Embung (Reservoir Basin)	The basin is at least 10 m long, 5 m wide and 2.5 m deep. Estimated static volume: 125 m^3/unit.	Provides a source of raw water for domestic use, irrigation in the dry season and a source of drinking water for large livestock.	Collect rainwater and surface runoff in areas around water sources such as springs, ditches and/or tributaries.	Cultivation, Settlement and Forest Areas.
Biopore	Holes (pores) in the soil made by biological bodies, such as earthworms, mice, marmots, hedgehogs, crickets, ants/termites, dead and rotting roots, including holes made using earth drills by humans. Man-made biopores are generally 10 cm in diameter with a depth of up to 100 cm. Estimated static volume: 16 liters/unit.	Distribution of oxygen and water in the soil	Increase infiltration capacity, store and absorb rainwater.	Cultivation and Settlement/urban Areas.
Kombipor	The composting technique uses biopore holes. Into the biopore hole with a diameter of 10 cm, a depth of up to 100 cm with an estimated volume of 16 liters/unit, plant remains and/or household organic waste are placed. The ability to absorb water is at least 2 times greater than the static volume to 32 liters/unit.	Distribution of oxygen and water as well as a source of soil nutrients and controlling household waste	Increase infiltration capacity, store and absorb rainwater.	Cultivation and Settlement/urban Areas.
Route Well	Drilling holes into the ground until it reaches the aquifer layer (underground water) with a diameter of 6 inches, the depth depends on the aquifer layer (generally using a 6 inch diameter paralon pipe, 6 m long with a hole perforated in the pipe wall). Estimated static volume: 110 m^3/unit.	Reduce/control flooding when it rains	Passing runoff water into the aquifer	Settlements and urban areas

Sources: Rauf et al. 2013; Rauf et al. 2016; Rauf et al. 2020.

Table 2. Watershed management using vegetative methods in efforts to protect water sources control erosion and sedimentation and landslides in North Sumatra

Type of Plant	Regional Suitability and Implementation Directions	Conservation Function	Production and Economic Functions
Aren (*Arenga pinnata*)	Lowland to highland (50-1500 m above sea level), sloping land, river/lake banks, spring recharge, critical land, rocky areas, former surface mines.	Strengthening the grip of soil and rocks, rehabilitating critical land, absorbing and retaining rainwater, agroforestry systems	Nira, palm fruit, palm fiber, sugar industry, bioethanol, medicinal raw materials
Tamarind (*Tamarindus indica L.*)	Lowland to medium land (50-700 m.a.s.l.), road banks, urban forests	Absorbs pollution (CO_2)	Spices and cooking spices, food and medicine industry
Bamboo (*Bambusa sp.*)	Lowland to highland (50-1500 m above sea level), sloping land, river/lake borders, recharged springs.	Strengthening the grip of soil and rocks, rehabilitating critical land, absorbing and retaining rainwater, agroforestry systems	Furniture, building materials, crafts/home industry, vegetables (bamboo shoots).
Weeping fig or Benjamin tree (Ficus benjamina)	Lowland to highland (50-1500 m above sea level), sloping land, river/lake borders, recharged springs.	Strengthening soil and rock grip, rehabilitation of critical land, absorbing and retaining rainwater, absorbing air pollution (CO^2), agroforestry system.	Components of city parks, ecotourism areas.
Bengal almond/Indian almond/Malabar almond/Singapore almond/Tropical almond/Sea almond/Beach almond/Talisay tree/Umbrella tree (Terminalia catappa L.)	Lowland to mediumland (25-700 m above sea level), sloping land, river/lake banks, critical dry land.	Pioneer trees on critical land	Raw materials for medicines and pesticides, furniture and building materials
Mahogany tree (*Swietenia mahagoni* (L.) Jacq.)	Lowland to mediumland (25-700 m above sea level), road banks, urban forests.	Absorbs pollution (CO_2)	Raw materials for medicines, pesticides, furniture materials and building materials

Type of Plant	Regional Suitability and Implementation Directions	Conservation Function	Production and Economic Functions
Fiji/Longan/Matoa (*Pometia pinnata*)	Lowland to highland (50-1500 m above sea level), sloping land, river/lake borders, recharged springs.	Strengthening river/lake/cliff banks; Agroforestry system components	Commercial fruit
Pakam (*Pometia pinnata*)	Medium to highland (500-1500 m above sea level), sloping land, river/lake borders, recharged springs.	Strengthening cliff and river banks	Wildlife and Fish Food Sources
Alstonia beatricis/Pulai/Pule (*Alstonia scholaris*)	Lowland to mediumlaind (50-700 m.a.s.l.), road and park borders, river/lake borders, recharge springs.	CO_2 absorber, water absorber and storage, river bank reinforcement.	Furniture materials, household crafts, medicinal raw materials.
Silk tree/Chinese albizia *(Albizia falcataria)*	Mediumland to highland (500-1500 m above sea level), sloping land, critical dry land.	Pioneer trees on critical land	Animal feed, building materials
Simpur/Sempur (*Dillenia indica*)	Lowland to mediumland (50-700 m above sea level), sloping land, river/lake banks and borders, recharged springs.	Reinforcement of river/cliff banks	Confectioned foods and medicinal raw materials
Rain Tree (*Samanea saman*)	Lowland to medium land (25-700 m.a.s.l.), road banks, urban forests.	Absorbs pollution (CO^2) and water, cools the garden	Furniture and craft materials

Sources: Rauf, 2015; Rauf et al. 2016; Rauf, 2017.

Table 3. Involvement of parties in several outreach activities, studies, planning and realization of watershed management programs in North Sumatra

Programs/Activities	Stakeholders Involved	Location	Execution Time
Socialization of Integrated Watershed Management in each area (cultivation, settlements and forests)	• BAPPEDA Sumatera Utara • BPDAS Wampu Sei Ular • BPDAS Asahan Barumun • Plantation Companies • Regency/City Government in North Sumatra • Farmer Groups • USU Academics	All areas of North Sumatra province	2007 to present
Preparation of an Integrated Watershed Management Plan	• BAPPEDA Sumatera Utara • BPDAS Wampu Sei Ular • BPDAS Asahan Barumun • USU Academics	All areas of North Sumatra province	2010 to present (revision)
Reclamation of ex-mining land C using an agroforestry system	• Forum DAS Wampu Sumatera Utara • Kelompok Konservasi Kukam Hijau • BPDAS WAmpu Sei Ular • USU Academics	Kukam Riverbank, Kuala Mencirim Village, Sei Bingei District, Langkat Regency	2010 to present
Multifunction and role of Rivers; Maintaining Springs, and Planting Trees to Strengthen River Banks	• Klinik Reboisasi • Forum DAS Sumatera Utara • Komunitas Pemuda Peduli Lingkungan (KOPPLING)	Upstream areas of the Deli watershed, Belawan watershed, Wampu watershed	2020 to present
Construction of Infiltration Well Prototype	• USAID IUWASH Plus • LPPM USU • Forum DAS Sumut • PDAM Tirtanadi Sumatera Utara • PDAM Tirtauli Pematang Siantar	Tirtanadi Springs Recharge Area, District. Sibolangit, Deli Serdang Regency and the Tirtauli Springs Recharge Area in Nagahuta District, Simalungun Regency	2020-2021

Programs/Activities	Stakeholders Involved	Location	Execution Time
Preparation of the Green Growth Plan (GGP) for North Sumatra Province	• BAPPELITBANG Prov. Sumatera Utara • BAPPENAS • USU Academics	All areas of North Sumatra province	2021-2022
Implementation of Critical Land Conservation Study Results in the Upper Asahan Watershed	• PJT-I • LPPM USU • PT Bajradaya Sentranusa • Toba Regency	Silaen District, Habinsaran, Siantar Norumonda, Toba Regency	2020-2021
Preparation of the Master Plan for the Toba Lake Water Catchment Area (DTA) Conservation Program and Watershed Areas Asahan River.	• INALUM • PT Surveyor Indonesia • USU Academics • KPH III Kisaran • KPH IV Balige • KPH XIII Dolok Sanggul	Lake Toba Catchment Area (Karo, Simalungun, Samosir, Toba, North Tapanuli, Humbang Hasundutan, Dairi Regencies)	2021-2022
Preparation of the Integrated Agricultural Area Masterplan for Deli Serdang Regency	• Deli Serdang Regency • Lapakbuah.com • USU Academics	STM Hulu District, Deli Serdang Regency	2022
Preparation of the Simalungun Regency Integrated Agricultural Area Masterplan	• Simalungun Regency • Lapakbuah.com • USU Academics	Dolok Silau District, Simalungun Regency	2023
Reclamation of ex-mining land C using Aren and Sengon trees	• Forum DAS Sumatera Utara • Pesantren Al-Hidayah Kutalimbaru Deli Serdang • Universitas Sumatera Utara • DPW PERWAKU	Sei Mencirim Village, Kutalimbari District, Deli Serdang Regency	2022 to present
Construction of Reservoirs/Medium Reservoirs and Preservation of Springs	• BPDAS Wampu Sei Ular • Farum DAS Belawan • Klinik Reboisasi • USU Academics and USM Indonesia • KSM Pondok Miri Asri	Buffer Area Semayang River Springs, Dusun X Medan Krio Village Sunggal District Deli Serdang Regency	2022 to present

Table 3. (Continued)

Programs/Activities	Stakeholders Involved	Location	Execution Time
Land Suitability Study for Optimizing Sustainable Agriculture in the PT Agincourt Resources Mine Circle	• PT. Agincourt Resources • PT. Surveyor Indonesia • USU Academics	Batang Toru and Muara Batang Toru Districts. South Tapanuli	2023-2024
Recommendations for Management of Forest Areas with Special Purposes (KHDTK)	• LPPM USU • Institut Teknologi DEL • KPH XIII Dolok Sanggul	Pollung District, Humbang Hasundutan Regency	2022-2023
Water Resources Vulnerability Assessment Wampu Watershed, Sei Bingei Sub-Watershed (for Mebesar Regional SPAM) Due to Climate Change	• USAID IUWAS Tangguh • PT. Fitrah Alam Indonesia • USU Academics	Sei Bingei District, Langkat Regency and South Binjai District, Binjai City	2022-2023
Development of Agribusiness Based on Sugar Palm and Sheep Farming to Support Islamic Boarding School Operations	• Pesantren Al-Hidayah • DPW PERWAKU • USU Academics	Sei Mencirim Village, Kutalimbaru District, Deli Serdang Regency	2022 to present
Ecotourism Development	• Jendela Bambu Leuser (JBL) • Klinik Reboisasi • Forest Farmers Group • Forum DAS Sumatera Utara	Upstream areas of the Deli watershed, Belawan watershed, Wampu watershed	2020 to present
Carbon trading and climate conscious village	• USU • UTM Malaysia • Kuala Bali Farm Serdang Bedagai	Kuala Bali Village, Sidodadi District, Serdang Bedagai Regency (Central Area of the Ular Watershed)	2024

Land Management Based on Watershed Management

The key to successful watershed management depends on land management and utilization in all areas that can maintain their productivity, without land degradation occurring in the form of erosion, landslides, floods and/or drought. For this reason, the basic principles of watershed management-based land management and utilization can include: (1) Increasing the soil's water absorption capacity (Infiltration capacity); (2) Harvesting rainwater (Water Harvesting); (3) Temporarily restrain rainwater from being directly discharged into drainage channels; (4) Management of lakes becomes a business area based on water management; (5) Repair/maintenance of water bodies so that they can still hold and distribute water at their maximum capacity; (6) Applies to all areas, both cultivation areas and residential areas, as well as forest areas.

Various land use technologies based on watershed management in North Sumatra have been implemented and continue to be socialized to be implemented in all areas or land use designations. In cultivation areas, various land use technologies that have been implemented can be seen in Table 1.

Apart from the mechanical techniques (methods) mentioned above, vegetative techniques (methods) based on watershed management, especially in reducing the rate of erosion and preventing landslides, have been widely used in North Sumatra, with location specifications and suitability of plants and their intended use, as presented in Table 2.

The realization of all the watershed management activities and programs mentioned above (Table 1 and Table 2) is carried out by the North Sumatra Province Watershed Management Coordination Forum in collaboration with various parties, both central and regional governments, universities, non-governmental organizations and communities as well as national and international NGOs., as can be seen in Table 3.

Conclusion

1. Integrated watershed management can control land degradation in the form of erosion, landslides, floods and/or drought, especially in wet tropical areas, such as in North Sumatra, Indonesia.

2. Land use technology based on watershed management is primarily aimed at maintaining land productivity in the sense that there is no land degradation in the form of erosion, landslides, floods and/or drought in all areas.
3. Involvement of stakeholders is a key factor in the success of watershed management for various purposes in all regions.

Disclaimer

None.

References

Arsyad, S. 1989. *Konservasi Tanah dan Air*. Institut Pertanian Bogor. Bogor.

Arsyad, S., Pryanto, A., Nasoetion, L. I. 1985. *Pengembangan Daerah Aliran Sungai.* Institut Pertanian Bogor. Bogor.

Asdak, C. 2002. *Hidrologi dan Pengelolaan Daerah Aliran Sungai:* Gajah Mada Univesity Press. Yogyakarta.

BPDAS Asahan Barumun. 2006. Karakteristik DAS Asahan. *BPDAS Asahan Barumun, Direktorat Jenderal Rehabilitasi Lahan Dan Perhutanan Sosial Departemen Kehutanan.* Pematang Siantar.

BPDAS Asahan Barumun. 2009. *Karakteristik DAS DTA Toba.* BPDAS Asahan Barumun, Direktorat Jenderal Rehabilitasi Lahan Dan Perhutanan Sosial Departemen Kehutanan. Pematang Siantar.

BPDAS Asahan Barumun. 2009. *Rencana Teknik Rehabilitasi Hutan dan Lahan DAS (RTk-RHL DAS).* BPDAS Asahan Barumun, Direktorat Jenderal Rehabilitasi Lahan Dan Perhutanan Sosial Departemen Kehutanan. Pematang Siantar.

Departemen Kehutanan RI. 1996. *Pedoman Identifikasi Karakteristrik DAS.* Direktorat Jenderal Reboisasi dan Rehabilitasi Lahan. Jakarta.

Departemen Kehutanan RI. *Balai Pengelolaan Daerah Aliran Sungai (BPDAS) Citarum – Ciliwung. Direktorat Jenderal Rehabilitasi Lahan Dan Perhutanan Sosial Departemen Kehutanan. 2008*. Pengelolaan Das Terpadu Das Citarum. Jakarta.

Departemen Kehutanan RI. BPDAS Asahan Barumun. 2003. *Rencana Teknik Lapangan Rehabilitasi Lahan dan Konservasi Tanah (RTL-RLKT)* DAS Asahan. BPDAS Asahan Barumun, Direktorat Jenderal Rehabilitasi Lahan Dan Perhutanan Sosial. Pematang Siantar.

Dunne, T., Leopold, L. B. 1978. *Water in Environmental Planning.* W. H. Freeman and co. New York.

Rahmawaty, Rauf, A. 2019. *Menelusuri DAS Singkil dari Hulu ke Hilir*. USU Press, Medan 2020.

Rauf, A., Sianturi, H., Rahmawaty, Y., Hidayat, B., Slamet. 2013. *Pengelolaan Daerah Aliran Sungai; Suatu Rencana Pengelolaan Terpadu DAS Asahan Toba*. USU Press, Medan.

Rauf, A., Rahmawaty, Sofyan, A. 2016. *Teknologi Pemanfaatan Lahan Berbasis Pengelolaan Daerah Aliran Sungai (DAS)*. USU Press, Medan.

Rauf, A. 2017. Sistem Agroforestry; Upaya Pemberdayaan Lahan Secara Berkelanjutan; Edisi Revisi. USU Press Medan.

Rauf, A., Nasution, H. B. H., Supriadi, H., Syafrizal. 2020. *Building Capacity of Local Government in Implementing KKMA Action Plans Through Infiltration Ponds in North Sumatra*. Laporan Akhir; Kerjasama IUWASH-PLUS USAID dan LPPM USU, Medan.

Rauf, A., Samsuri, Rahmawaty, Slamet, B., Rahmanta, Supriadi. *2021. Kajian Implementasi Hasil Studi Lahan Kritis Daerah Aliran Sungai Asahan Hulu*. Laporan Akhir; Kerjasama Perum Jasa Tirta I dan LPPM USU, Medan.

Rauf, A., Samsuri, Rahmawaty, Slamet, B., Rahmanta, M., Harahap, M. 2022. *Inventarisasi Lahan Kawasan Hutan dengan Tujuan Khusus (KHDTK) IT-DEL*. Laporan Akhir; Kerjasama Institut Teknologi-DEL dan LPPM USU, Medan.

Rauf, A., Rahmawaty. 2019. *Ekspedisi DAS Wampu Provinsi Sumatera Utara*. USU Press Medan, 2020.

Rauf, A. 2015. *Multifungsi Pohon Pakam* (*Pometia spp.*); Fungsi Konservasi Sumberdaya Alam, Ekonomi, dan Obat. https://araufismail.blogspot.com/2015/04/multi-fungsi-pohon-pakam-matoa-batak.html.

Chapter 7

The Model of Liability for Compensation for Environmental Pollution or Destruction in Indonesia

Suhaidi[*]
and Sunarmi
Law Faculty, Universitas Sumatera Utara, Medan, Indonesia

Abstract

The discussion on the environment and its status as a legal subject to ensure the sustainability of a good and healthy environment is a relevant issue today. This article discusses responsibility and compensation for polluters who harm the environment. There are three forms of civil liability for environmental pollution. First, "liability based on fault." Indonesia recognizes this responsibility from unlawful acts as stated in Article 1365 of the Civil Code. Second, "strict liability." This liability model is based on the "lex specialis" of unlawful acts. The specificity referred to here is in the Pollutant Category, which is every person whose actions, business and/or activities use dangerous and toxic materials (*Bahan Berbahaya dan Beracun*-B3), produce and/or manage B3 waste, and/or pose a serious threat to the environment. Along with developments in handling environmental pollution, there is another form of liability in the case of marine pollution caused by an oil spill on Braer ship. The International Convention on Civil Liability for Oil Pollution Damage (CLC) has legalized this and stated that ship owners can be subject to absolute liability if the accident that occurs is truly the fault of the ship owner,

[*] Corresponding Author's Email: suhaidi@usu.ac.id.

In: Natural Resources, Environment, and Food and Energy Security …
Editors: Mohammad Basyuni and Severino Salmo III
ISBN: 979-8-89530-283-5

with a limit of compensation that can be paid in accordance with the ship insurance policy. The 1969 CLC regulations can be implemented in Indonesian legal regulations to ensure the sustainability of the marine environment.

Keywords: fault, marine, environmental, pollution, absolute liability, strict liability, Indonesia

Introduction

Personification is applied in the enforcement of environmental laws. Some studies on environmental law enforcement explain how the environment can become a legal subject recognized as having rights (Usman, 2018). In Indonesia, the environmental is also considerd a rights holder, ensuring its protection to guarantee human survival (Usman, 2018).

The responsibility for environmental management lies and rests with the government, which has an impact on institutions, and it is the authority of the government to manage the environment as a part of administrative law (Febriyanti, Aini, Resta, and P.K.P, 2021). Environmental regulations in Indonesia are regulated by Law Number 32/2009 concerning Environmental Protection and Management (UUPPLH). Law Number 6/2023 concerning the Determination of Government Regulations instead of Law Number 2/2022 concerning Job Creation Becoming Law (stated as Job Creation Law 2023) has the character of an "umbrella act" for the environmental laws (Febriayanti et al. 2021).

The UUPPLH and Job Creation Law 2023 were enacted to provide legal protection for environmental management. Furthermore, preventing environmental damage or pollution is the first step towards ensuring environmental protection since deteriorating changes in the quality of the environment threaten the continuity of human life and that of other living organisms (Febriyanti et al. 2021).

It is difficult to predict environmental changes caused by pollution and/or environmental destruction, especially in terms of biological wealth. It requires continuous monitoring of the environment. The *Group of Experts on the Scientific of Marine Pollution* (GESAMP) mentions the following changes in the marine environment:

"The difficulty of monitoring biological change given its long-term and extensive geographic variation, and the problem of interpreting these observations in the light of the results of experimental exposures of a few species to a few known contaminants, are not easily resolved. Where changes are expected, for example, in areas receiving discharges, monitoring programmes may be initiated, but they involve an expensive and long-term commitment of resources; in addition, it is often charged that irreversible damage will ensue before the results of long-term programmes are accumulated and analyzed."

Liability for the Environment

Legal liability for the environment is known as civil liability (Riswanti, Martana, and Dananjaya, 2013). This means that law enforcement due to pollution and/or damage to the environment through compensation must aim to restore the original environment or must be given for the damage caused (Gold, 1991). Compensation may also be given to the affected communities whose livelihoods are disrupted (Suhaidi et al. 2024).

Concerned person and/or company is responsible for their respective business and/or activity that commits an unlawful act in the form of environmental pollution. It (they) is (are) obliged to pay compensation and/or take specific actions (DPR, 2009), based on the following:

a. Failure to fulfill contractual obligations as stated in Article 1243 of the Civil Code;
b. Unlawful Act as stated in Article 1365 of the Civil Code.

Unlawful Act as stated in point b is an accumulation of a principle known as "liability based on fault." Here, the victim of the pollution/plaintiff bears proof. Thus, the plaintiff receives compensation only if the victim proves that the polluter and/or environmental destroyer is at fault. A fault is an element that determines liability. If he/she/it is not proven guilty, no compensation is paid. Furthermore, every person whose actions, business, and/or activities use B3, produces and/or manages B3 waste, and/or poses a serious threat to the environment, is absolutely liable for losses arising from their business and/or activities (DPR, 2023). Absolute liability is also known as strict liability. In the Netherlands, this theory is known as *"risico aansprakelijkheid,"* and in France it is called *"theorie du risque cree"* (Lotulung, 1998).

Internationally, the principle of strict liability is known and used by many states, including Indonesia. The explanation of Article 22/33 UUPPLH describes the "strict liability" as the element of fault, however, it does not need to be proven by the plaintiff as a basis for payment of compensation. It is the *"lex specialis"* of unlawful acts in general which are based on fault. The principle of absolute liability in the context of the obligation to pay compensation must be viewed as a risk to every business and/or activity that causes danger or has the potential to cause dangerous impacts on a person/society and/or the environment.

The US Court determines the types of activities where the principle of strict liability can be applied as "ultra hazardous or abnormally dangerous activities." These must comply with the following criteria: (a) the level of risk is high, (b) the likelihood of harm being caused, (c) the ability to eliminate risks by implementing them carefully, (d) the extent to which the activity is not everyday in society, (e) the feasibility of carrying out activities at a particular location, and (f) the value or benefits of the activity for the community.

Loss of Economic Value as an Impact of Environmental Pollution and/or Destruction

Natural resources in the environment provide goods and services as well as economic benefits (Djajadiningrat, 1996). Plants, animals, and fish species live in the environment and are part of a country's economy (Gore, 1995). The presence of pollutants/contaminants in the environment has several negative consequences. The environment will be affected both naturally and economically, damaging the ecosystem of life on Earth (Hardjasoemantri, 2006) and harming food supply chains and living habitats which are sources of natural wealth for a country (Schiffman, 1996).

Because the environment naturally consists of land, water, and air, environmental pollution and/or destruction impact international commerce and investment (Percival and Alevizatos, 1997). For example, exports of CPO commodities, such as palm oil trees planted in forest areas and furniture commodities whose basic materials result from illegal logging processes in protected forest areas. Environmental losses can also occur at sea in the form of property damage, preventive measures, pure economic losses, and environmental damage (Guruswamy et al. 1994).

The loss of economic value in the environmental pollution compensation system is that if damage due to environmental pollution and/or destruction cannot be avoided, comparable compensation must be given to victims of environmental pollution and/or destruction (Gold, 1991). Calculating losses in economic value involves covering and comparing the conditions before and after environmental pollution and/or destruction had occurred. The amount of compensation is the amount of money that can be legally accounted for by the polluter and/or environmental destroyer (Kantaatmadja, 1981). Apart from the difficulty of amassing proof, it is extremely difficult to assess this amount accurately (Kantaatmadja, 2981). Other categories can be submitted as the basis for a lawsuit and determined based on facts that can be accounted for accompanied by supporting documentation (IOPC, 2000).

For example, in claims for compensation for marine environmental pollution caused by oil from tanker accidents, the International Maritime Organization (IMO) has created a guideline that mentions the need for the following data and information (IMO, 1988): (1) the name and address of the claimant or his/her representative; (2) the ship profile denoted in the accident; (3) time, place, and detailed data regarding the accident; (4) data on pollution prevention and cleanup, including types of damage to the marine environment; and (5) the extent of the demands.

The losses that can be claimed are (Kantaatmadja, 1981): (1) losses arising from the costs of cleaning up pollutants; (2) socioeconomic losses, including losses in the production, trade, and social sectors; and (3) ecological losses, which include monitoring of coastal damage operations, such as the rehabilitation of polluted areas in the form of restoring them to their original condition and losses due to reduced production during the environmental regeneration period, are still legally contested (Kantaatmadja, 1981).

The Model of Compensation System According to Environmental Law in Indonesia

The regulations of environmental law in Indonesia related to civil liability, namely "liability based on fault" (Article 1365 of the Civil Code) and strict liability (Article 88 of UUPLH) have been discussed so far. In addition to regulating these two compensation system models, the 2009 UUPLH and Job Creation Law-2023 also regulate environmental disputes. There are two ways to resolve these disputes: (1) settlement of disputes through out-of-court channels to reach an agreement regarding (a) the form and amount of compensation, (b) recovery measures resulting from pollution and/or damage, (c) actions to ensure that the impact of pollution and/or destruction is not repeated and/or (d) actions to prevent adverse environmental impacts. The ideas in both UUPPLH and Job Creation Law-2023 allow the resolution of environmental disputes outside the court, which is unrestricted, impartial, the choice of the parties, and voluntary based on the principle of deliberation; (2) Dispute resolution through the courts. It is believed that environmental dispute resolution will be more effective in obtaining a sense of justice and will be accepted by the parties to the dispute quickly and at a low cost.

Parties who choose to settle disputes out-of-court allow the use of a third party as a mediator and/or arbitrator when there are no environmental crimes. The third party can be an arbitrator, mediator, or another third party. Similarly, the community can establish an institution that provides environmental dispute resolution services. The central government can facilitate the establishment of an institution that provides environmental dispute-resolution services (DPR, 2009).

Accountability for the compensation of environmental pollution occurs on land and in sea (Suhaidi, Rosmalinda, and Arif, 2021). Generally, the principle used is that of strict liability (IMO, 1975). As discussed above, strict liability implies that if oil causes damage, the tanker owner is liable for that damage, even if the tanker owner was not negligent. Strict liability is liability without fault, and is often the standard of liability used by law regarding harm resulting from participating in dangerous activities such as using dynamite or toxic substances (Firestone and Reed, 1993). The Civil Liability Convention (CLC) is an international convention concerning civil liability for losses resulting from oil pollution. This convention resulted from the IMO conference, which was held to ratify regulations regarding responsibility for damage to the marine environment due to oil spills. Indonesia, as an archipelagic country that plans to become the world's maritime center, has paid special attention to the marine environment, considering Indonesia's existence as a sea crossing that is at risk of oil spills (Suhaidi et al. 2018).

Compensation cannot be claimed from the ship owner if the accident is caused by: (a) war, armed conflict, civil war, rebellion, or natural disaster;

(b) acts or negligence of third parties to cause the accident; and (c) negligence or wrongful actions of the government or officials responsible for maintaining lighthouses, signs, and other navigational aids. By the principles used, namely the principle of strict liability, the limit of compensation that can be paid is by ship insurance (IMO, 1975).

However, the shipowner can be subject to absolute liability if the accident that occurs is indeed the fault of the shipowner (IMO, 1975). Thus, there is another form of civil liability, namely absolute liability, if the shipowner is conclusively at fault in the event of a tanker accident. Absolute liability can be included in Indonesia's environmental laws to accommodate the resolution of cases of environmental pollution and/or damage.

Conclusion

The importance of environmental protection and legal liability needs to be established for violators who damage the environment. There are three forms of civil liability recognized in Indonesian environmental law: first, "fault-based liability" provided for in Article 1365 of the Civil Code; second, "strict liability" applied to individuals or entities that use hazardous and toxic substances (B3) and that may pose a serious threat to the environment; and third, special liability related to marine pollution from oil spills, provided for by the International Convention on Civil Liability for Oil Pollution Damage (CLC). Through this legal framework, it is hoped that an effective compensation mechanism can be created to restore environmental damage and provide protection for affected communities. This research also emphasizes the need for ongoing monitoring and strict law enforcement to prevent further damage and ensure the sustainability of a healthy environment for future generations.

Disclaimer

This article is a stimulated information regarding the Liability Model for Compensation of Environmental Pollution and/or Damage in Indonesia. Readers are welcome to enhance the reading material from other sources regarding the liability for losses due to environmental pollution and/or destruction, both inside and outside Indonesia.

References

Djajadiningrat, S. T. 1996. [*Utilization of Economic Instruments for Environmental Management*]. Jakarta: Kantor Menteri LH & Bapedal.

DPR. [*Law Number 32 Year 2009 on Environmental Management and Protection*]. Pub. L. No. 32. 2009. Indonesia.

DPR. [*Law Number 6 of 2023 on Job Creation*]. Pub. L. No. 6. 2023. Indonesia.

Firestone, D. B. and Reed, F. C. 1993. [*Environmental Law for Non-Lawyers*]. California: Soro Press.

Gold, E. Compensation in Oil Pollution Damage [International Seminar Summary] in *Bandung: The ICLOS-NILOS Seminar Law of the Sea in 1990'S: Offshore Development in Indonesia.* 1991 p. 261.

Gore, A. 1995. Marine degradation From Land-based Activities: A global Concern. *U.S Department of State Dispatch 6*(46): 2.

Guruswamy, L. D., Palmer, G. W., Weston, B. H. 1994. *International Environment Law and World Order: A Problem-Oriented Coursebook.* St. Paul: West Publishing Company.

Hardjasoemantri, K. 2006. *Hukum Tata Lingkungan* (8th ed.). Yogyakarta: Gadjah Mada University Press.

IMO. *International Convention on Civil Liability for Oil Pollution Damage*. International Maritime Organization. 1975.

IMO. *Manual on Oil Polution, Section II Contingency Planning.* London: International Maritime Organization. 1998.

IOPC. *International Oil Pollution Compensation Fund Claim Manual.* London: Whitstable litho Ltd. 2000.

Kantaatmadja, K. 1981. [International Compensation for Oil Pollution at Sea]. Bandung: Alumni.

Lotulung, P. E. 1998. *Penegakan Hukum Lingkungan Dalam UU No. 23 Tahun1997 Ditinjau Dari Aspek Hukum Perdata.* Semarang.

Percival, R. V., and Alevizatos, D. C. 1997. *Law and the Environment: A Multidisciplinary Reader*. Philadelphia: Temple University Press.

Riswanti, A. R., Martana, N. A., and Dananjaya, I. N. S. 2013. [Strict Liability in Environmental Civil Law Enforcement in Indonesia]. *Kertha Wicara: Journal Ilmu Hukum*, 1(3), 1–5. Retrieved from https://ojs.unud.ac.id/index.php/kerthawicara/article/view/6100.

Schiffman, H. S. 1996. "The Protection of Whales in International Law: A Perspective for the Next Century." *Brooklyn Journal of International Law* 22(2):2. Retrieved from. ttps://brooklynworks.brooklaw.edu/bjil/vol22/iss2/1.

Suhaidi, Alhayyan, R., and Leviza, J. 2024. [Legal Understanding of the Compensation Model in Environmental Law in the Community in Pardomuan Ajibata Village, Toba Regency] *Abdimas: Jurnal Pengabdian Masyarakat* 5(1): 1398–1403.

Suhaidi, Rosmalinda, and Alhayyan, R. The Global Maritime Nexus: Challenges in Managing Marine Area in North Sumatra Province. *ICOSTEERR: International Conference of Science, Technology, Engineering, Environmental and Ramification Researches*, 2018, pp1663–1669. Medan: Scitepress Digital Library.

Suhaidi, Rosmalinda, and Arif. 2021. "Coastal Tourism: Marine Pollution and Lack of Local Government Efforts." *NVEO-Natural Volatiles & Essential Oils Journal 8*(4): 81–94. Retrieved from https://www.nveo.org/index.php/journal/article/view/147.

Usman, A. S. 2018. [The Environment as a Legal Subject: Redefining the Relationship between Human Rights and Environmental Rights in the Perspective of the Rule of Law]. *Legality: Jurnal Ilmiah Hukum 26*(1):1–16. Retrieved from https://ejournal.umm.ac.id/index.php/legality/article/view/6610.

Chapter 8

Renewable Energy Potential in Sumatra Utara Province: An Overview

Himsar Ambarita*
and Ilmi Abdullah
Department of Electrical Engineering, Faculty of Engineering,
Universitas Muhammadiyah Sumatera Utara, Indonesia

Abstract

Sumatera Utara Province has abundant renewable energy potency that can be developed. These include hydropower, geothermal, biomass, biogas, solar, and municipal solid waste. However, these potencies must be explored optimally, and technical and economic barriers must be considered. Several regulations have been enacted to accelerate the exploration of renewable energy. Optimal collaboration is required between all parties and policymakers in order to support the development of renewable energy projects. The aim of this study was to determine the priority sequence for renewable energy development, particularly in the Sumatera Utara Province. These considerations include geographic location, technology implementation, cost, and availability. To provide data for analysis, several statistical and study reports were reviewed. The results show that for Sumatra Utara Province, hydropower has the highest potency because it is supported by many rivers flowing from the mountains and has the most mature technology. Hydropower is followed by geothermal energy because of the location: along Bukit Barisan, an active mountain range with very large energy reserves. The nexts are biomass and biogas from the palm oil industry, as this province is known as the 4th largest palm oil

* Corresponding Author's Email: himsar@usu.ac.id.

In: Natural Resources, Environment, and Food and Energy Security …
Editors: Mohammad Basyuni and Severino Salmo III
ISBN: 979-8-89530-283-5

production province in Indonesia. Solar power, municipal solid waste, and wind power are at low levels of exploration. However, municipal solid waste energy has a very high potential to be explored, and it is expected to be a solution to the urban waste problem present.

Keywords: renewable energy, geography, hybrid, economic

Introduction

The geographical location of Sumatra Utara Province is between 1° – 4° North Latitude and 98° – 100° East Longitude, with an area of 182,414.25 km^2. The province comprises 60.5% oceans and 39.5% land. It is well known that this province has renewable energy reserves with great potential for development. In addition, the province has a series of active mountains, such as Mount Sibayak and Sinabung, with an altitude of 2,460 m above sea level, and Martimbang and Sorik Merapi, which have large geothermal energy potential. According to the Geological Agency (Geothermal Directorate of Indonesia) as of December 2019, the geothermal potential in Indonesia was 23.9 GigaWatt GW. Sumatra Utara, one of the provinces, has the largest geothermal energy potential, namely 1,857.00 MW that is spread across six regencies. These included Karo, Simalungun, North Tapanuli, South Tapanuli, Padang Lawas, and Mandailing Natal. In development planning, the government aims to increase the use of geothermal energy to 7,241.5 MW of the total potential (16.8 %) by 2025 (Mahmudah, 2018).

The rivers that flow from the Bukit Barisan Mountains, such as The Asahan, Silau, Barumun, Bilah, Batang Girls, Deli, Kualuh, Lau Tenges, Mencirim, Renun, Simpang Kanan, Toru, Wampu, Bekulap, Bingai and Bahorok Rivers, are also potential energy storage areas in Sumatra Utara Province. According to 2020 Central Statistics Agency data, another potential water resource is Lake Toba, which has an area of 112, 920 ha or 1.57% of the total area of Sumatra Utara and is a extremely large hydro site that supports the development of renewable energy in Sumatra Utara (Stocks et al. 2019). Sumatera Utara Province also has biomass potential that can be drawn from palm oil plantations with an area of 1,256,808 ha. This plantation can be divided into 505,882.84 ha (40%) is managed by private companies, 321,663.85 ha (26%) is managed by state plantation companies and the rest 429,261.31 ha is smallholder oil palm plantations. Based on data from the Central Statistics Agency, in 2020, the total production of Fresh

Fruit Bunches (FFB) reached 11,621,750.32 tonnes, managed by 32 integrated palm oil mills using biomass as fuel. From an electricity potential analysis conducted by (Nasution et al. 2014) on several palm oil mills in Sumatra Utara, it was concluded that a mill with a capacity of 30 tons of FFB per hour could generate 20–35 MW of electrical power. Not all biomass fuel plants are connected to the existing grid yet; therefore, they cannot carry out measurable exports or imports.

Not all of the present renewable energy potency from natural resources and plantations have been optimally utilized in the past. Indonesia has an ambitious target to decrease the utilization of fossil fuels by increasing the share of renewable energy. The target can be found in Government Regulation No. 79 of 2014, which states that the renewable energy share can reach 23% in 2023 and 31% in 2050 (KESDM, 2014). This study aims to examine the potential of natural resources that can be developed as a sustainable source of renewable energy in Sumatra, Utara, towards net-zero emissions by 2060.

Government ACT and Policy

The government of Indonesia has regulated renewable energy policies through Law Number 30, 2007 concerning Energy. This law requires the government to increase the provision of New Energy and Renewable Energy in order to reduce the gap between policy and regulation and to support the achievement of national energy mix targets (KESDM, 2007). This law requires governments to increase the provision of new and renewable energy sources. To achieve the announced energy mix target of 23% by 2025, there should be coordination and synergy between institutions and governments to encourage the development of renewable energy. Subsequently, the renewable energy roadmap must be clear, and there is a need to identify future opportunities and challenges, especially in the implementation phase of the 2020 NDC. The Indonesian government has issued several policies related to renewable energy. They are Presidential Regulation Number 4 of 2016 concerning the Utilization of Renewable Energy (KESDM, 2016), the Minister of Energy and Mineral Resources regulation Number 12 of 2017 concerning the Use of Rooftop Solar Power Generation Systems by Consumers of State Electricity Company (Persero) (KESDM, 2017), the Minister of Energy and Mineral Resources regulation Number 49 of 2018 concerning Power Plant Development Microhydro Electric Power (KESDM,

2018a), Regulation of the Minister of Energy and Mineral Resources Number 50 of 2018 concerning Development of Wind Power Plants (KESDM, 2018a), Regulation of the Minister of Energy and Mineral Resources Number 51 of 2018 concerning Development of Solar Power Plants (KESDM, 2018b), Regulation Minister of Energy and Mineral Resources Number 52 of 2018 concerning Development of Wave Power Plants (KESDM, 2018c), Regulation of the Minister of Energy and Mineral Resources Number 53 of 2018 concerning Development of Geothermal Power Plants (KESDM, 2020), Regulation of the Minister of Energy and Mineral Resources Number 12 of 2019 concerning the Development of Waste Power Plants (KESDM, 2019a), Regulation of the Minister of Energy and Mineral Resources Number 13 of 2019 concerning the Development of Biogas Power Plants (KESDM, 2019b. With these policies, it is hoped that the development of renewable energy in Indonesia will continue to improve and provide long-term benefits to the economic and environmental sectors. In addition, in 2022, the Indonesian government will issue Presidential Regulation Number 112 of 2022 concerning the Acceleration of Renewable Energy Development for the Supply of Electric Power (KESDM, 2022). This regulation aims to increase investment and accelerate the achievement of renewable energy mix targets in the national energy mix, in accordance with the National Energy Policy, as well as to reduce greenhouse gas emissions. These regulations were designed in order to accelerate the development of renewable sources in Indonesia. It is hoped that the development of renewable energy in Indonesia will continue to improve and provide long-term benefits for the economic and environmental sectors, and that there will be an increase in investment in the renewable energy sector, accelerating the implementation of renewable energy projects and supporting the achievement of renewable energy share in Indonesia. This is in line with the large economic potential of renewable energy in the Sumatra Utara Province, which, if utilized optimally, can become a significant economic resource and improve the welfare of people. Some of the things regulated in government policy include accelerating the development of electricity generation from renewable energy sources, establishing regulations that support the investment and development of renewable energy, adjusting fiscal and non-fiscal incentives to support the development of renewable energy, supporting geothermal development, and financing and guaranteeing facilities through state-owned enterprises.

Potential of Renewable Energy in Sumatra Utara

Hydropower Energy

Based on its geographical location, Sumatera Utara Province has a large hydropower resource potential for development as a renewable energy source. Hydropower resources that can be utilized as renewable energy sources in the Sumatra Utara Province include rivers, waterfalls, lakes, and dams. Several big rivers that flow in this region are the Batang Toru River, Asahan River, Bah Bolon River, and Wampu River which have the potential to be converted into electric power as a basic load for power lines. Some waterfalls are present in this province, such as the Sipiso-piso waterfall with a height of ± 120, and the Sigura-gura waterfall. The Sumatra Utara Province has a lake 100 km long (62 miles), 30 km wide (19 miles), and 508 m deep. This lake is located in the middle of the northern island of Sumatra and has a surface height of approximately 900 m above sea level. The lake stretches from 2.88°N 98.52°E to 2.35°N 99.1°E. Lake Toba is the largest lake in Indonesia as well as the largest volcanic lake in the world. This lake has a high potential for hydropower. The province also has several dams, including the Asahan, Sipansihaporas, and Lau Simeme dams. Some of the potential water resources that have been developed as renewable energy sources in Sumatra Utara province includes the hydroelectric power plant on the Asahan River, which is one of the largest hydroelectric power plants in Sumatra Utara with a capacity of 500 MW. There is also a hydroelectric power plant on the Batang Toru River and it has a capacity of 510 MW, one on the Bah Bolon River with a capacity of 126 MW, the Sipansihaporas Dam has a capacity of 192 MW, and the Asahan Dam has a capacity of 200 MW.

Hydropower with a smaller capacity named as microhydro power plants have also been developed, such as micro hydropower plant in Buluh Naman Village, Deli Serdang Regency with a capacity of 10 kW and has been operating since 2015 (Sakidiansyah, 2011), the micro hydropower plant in Manalu Dolok Village, District Parmonangan, North Tapanuli Regency has a capacity of 1.2 MW and was completed in 2021, the micro hydropower plant on the Peipei River, Dairi Regency has a capacity of 1.5 MW and has been operating since 2017, the micro hydropower plant in Aek Sibundong, North Tapanuli Regency has a capacity of 9 MW and has been operating since 2022, the Sei Wampu 1 Maryke micro hydropower plant in Langkat Regency has a capacity of 2 × 6 MW and has been operating since 2018, the Lau Gunung Micro hydropower plant, Tanah Pinem District Dairi has a capacity of 2×7.5 MW. These facts indicate that hydropower has undergone significant development. However, in comparison to the real potency, the developed power is still far from optimum.

Geothermal Energy

Indonesia is known as a country with the biggest potency of geothermal energy in the world. The exploration plan for the Indonesian geothermal field during 2020 – 2024 is shown in Figure 1. Sumatra Utara has the largest geothermal potential in Indonesia, with three geothermal blocks located in Sibayak, Sarulla, and Sorik Merapi. In the period–1993-1998, Unocal collaborated with PT. Pertamina explored geothermal resources in the Sarulla Block and discovered three geothermal sources, namely Sibualbuali, Silangkitang, and Namora-I-Langit, with a capacity of 330 MWe over 30 years. The geothermal working area of the Sarulla Block is 4.37 km^2 (Ompusunggu et al. 2017). The Sarrula geothermal is known as a geothermal power plant project with the largest single power purchase agreement (PPA) in the world. Construction started in 2014 using a private developer scheme (independent power producer/IPP). This project has been operating in Units I, II, and III, with a total capacity of 330 MW, since 2018. According to a document by the Ministry of Energy and Mineral Resources, geothermal energy in Sumatra Utara Province has eight additional sites being developed, namely the Sorik Marapi-Roburan-Sampuraga Mining Working Area (WKP), WKP Sipoholon Ria-ria, WKP Simbolon Samosir, WKP Mount Sibual Buali, Dolok Marawa Geothermal, Pincurak, and Sibubuhan Geothermal. The Sorik-Marapi geothermal system was developed in several stages. The first stage in 2017 was 20 MW, in 2018, 20 MW, in 2019, 40 MW and in 2026 it is 80 MW. However, until 2019, the Sorik Merapi geothermal system remained in the exploratory stage. The geothermal potential of Sipoholon Ria-ria was built at two sites, namely Helatoba, with speculative resources of 25 MW, and Sipoholon Ria-Ria, with suspected resources of 147 MW. The geothermal potential at this location is planned to develop by 2023, with an expansion of 10 MW × 2. The geothermal potential of Simbolon Samosir is located in the Samosir, Taput, Humbang Hasundutan, and Dairi districts, with estimated reserves of 150 MW. Development is planned for 2025 in two 55 MW stages. The geothermal potential of Mount Sibayak-Mount Sinabung spreads across three points: Beras Tepu, Lau Debuk-Debuk, and Marike. The site of Beras Tepu has hypothetical resources of 34 MW, suspected resources of 15 MW, possible resources of 20 MW and speculative resources of 20 MW. In Lau Debuk-

Debuk/Marike, there are speculative resources of 25 MW. This WKP is planned to be managed by Pertamina Geothermal Energy with a development capacity of 12 MW. The Pincurak geothermal potential in the Mandailing Natal Regency has speculative resources of 50 MW. In addition, the geothermal potential of South Sibubuhan Tapanuli has speculative resources of 100 MW. Based on the literature, geothermal energy systems are environmentally friendly and clean when compared with conventional energy systems. However, this technology cannot be categorized as free of environmental impacts (Bayer et al. 2013; Paulillo et al. 2019). Geothermal energy systems have shown encouraging substantial progress in terms of technical research and development, both in a hybrid configuration and as a standalone and environmentally friendly system (Assad et al. 2021). Therefore, it is feasible to develop them as a base load supplier for grids, and they can be hybridized with other technologies such as wind, hydro, and biomass (Li et al. 2020).

Biomass Energy

Indonesia is the largest palm oil producer worldwide. In Indonesia, the Sumatra Utara Province is the 4th biggest palm oil producer. According to a report from the Executive Director of the Palm Oil Agribusiness Strategic Institute (PASPI), in 2021, palm oil production in Sumatra Utara province reach 5.3 thousand tons. In the Indonesian economy, the Palm Oil industry is a strategic industry that contributes the largest amount of foreign exchange from the non-oil and gas sectors. In 2016, the Palm Oil Plantation Management Agency (BPDP Sawit) recorded that Indonesian palm oil exports reached 25.7 million tons, equivalent to USD 17.8 billion. The development of the Indonesian palm oil industry continues to increase, and Indonesia has become the largest Crude Palm Oil (CPO)-producing country in the world, with a total production of 32 million tons, or approximately 46.6% of the world's total CPO production. Global market demand for CPO continues to increase. According to global statistics, the estimated world demand for CPO in 2020 was 95.7 million tons, and it will continue to increase in the future. These facts show that the Indonesian palm oil industry has the potential to continue developing to meet the world's CPO needs. It can be said that Indonesia will play an important role in the world Palm Oil industry and the related industries.

Apart from increasing production and bright market projections for the palm oil industry, the other hand palm oil industry can cause problems such as an increase in the volume of waste and emissions. The issue of solid waste in the palm oil industry was scrutinized. Solid waste is utilized in fuel-to-power boilers (Aziz et al. 2017). Several types of solid wastes from the palm oil industry, such as empty fruit bunches, shells, and fibers, can be converted into energy using cogeneration systems. Palm oil processing produces solid, liquid, and gas wastes (Chyou et al. 2020). Solid waste included fresh fruit bunches (FFB), empty fruit bunches (EFB), mesocarp fiber (MF), palm shells (PKS), and palm kernel meal (PKM). Processing one ton of fresh fruit bunches from palm oil produces waste in the form of empty palm fruit bunches (up to 23% or 230 kg), shell waste (up to 6.5% or 65 kg), wet decanter solid (palm mud) (4% or 40 kg), fiber (13% or 130 kg), and liquid waste (up to 50% (Pranoto et al. 2013; Erivianto, 2018). The palm oil plantation area in Sumatra Utara Province can be divided into three categories: private plantation companies, government-owned companies, and smallholder farmers. Based on data from the Central Statistics Agency for 2020, it was recorded that the area of palm oil plantations in Sumatra Utara Province reached 1,256,808 ha, of which 505,882.84 ha or approximately 40% were managed by private plantation companies, around 26% or the equivalent of 321,663.85 ha were managed by government-owned plantation companies, and as much as 34%, or equivalent to 429,261.31 ha, were managed by smallholder farmers. Palm oil production in Sumatra Utara Province continued to increase from 5,119,497 tons in 2017 to 5,928,612 tons in 2021.

This production is related to the significance of solid waste. Several previous studies on the potency of palm oil mill solid waste for electricity production have been reported. For Sumatera Utara Province cases, a palm oil mill with a capacity of 30 tons of FFB/hour could produce 20 MW of electric power (Nazaruddin et al. 2014; Parinduri, 2016). At the same time, the palm oil mill also consumes a small amount of electricity to power the process. For a palm oil mill with a capacity of 40 t FFB/h, the required electricity was only 502.4 kW (Aditya, 2020). A more detailed study showed that every solid waste can produce electricity, using the following estimation: If it is used as fuel for the steam power plant, empty fruit bunches can generate 1.16 MW of electricity for each ton, and 1.10 MW from each ton of fiber and 1.16 MW from each ton of shell. Using these fuels, a steam power plant is expected to operate with a capacity factor of 83.6% (Putra et al. 2016; Khwee, 2018).

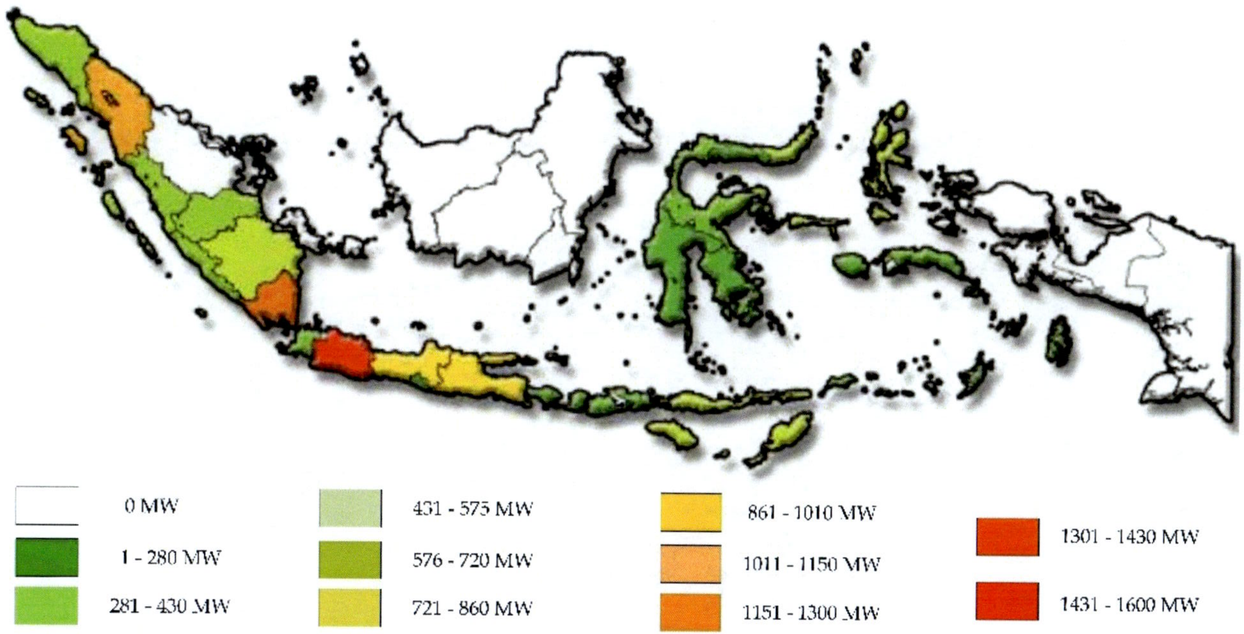

Figure 1. Exploration plan of geothermal potency in Indonesia.

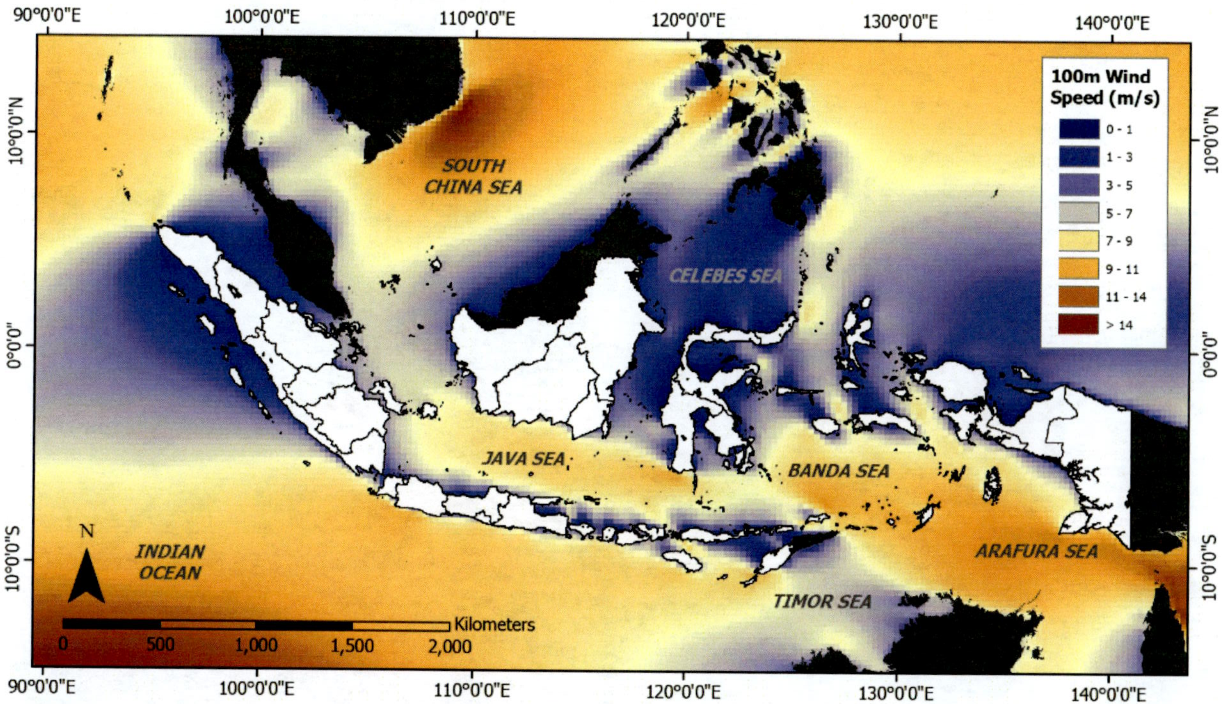

Figure 2. Wind Speeds in Indonesia at 100m (2013-2023). Data: ECMWF ERAS5.

From an economic perspective, biomass generation costs are in the range of 0.03-0.07 USD/kg and electricity production costs are in the range of 0.09-0.16 USD/kWh. This production cost is lower in comparison to oil-fueled diesel engines which are commonly used in remote areas in Indonesia (Susanto et al. 2018; Fujita et al. 2019). The electricity production costs obtained from the calculations based on the interest rates of 6%, 9%, and 12% were Rp. 572.67 Rp/kwh, Rp. 680.67 Rp/kwh and 814.32 Rp/kwh (Fujita et al. 2019). The effective use of waste can also supply future fuels to replace fossil fuels in power plants [30]. In terms of emissions, it was reported that greenhouse gas (GHG) emissions from the electricity production of empty bunch palm oil mills were only 0.15 kg $CO_{2\text{-eq}}$ per kWh (Danish and Ulucak, 2020).

By 2030, the production of solid waste from the palm oil industry will comprise 54 million tons of EFB, 31 million tons of MF, 15 million tons of PKS, 130 million tons of palm oil mill effluent (POME), 115 million tons of palm fronds and 59.7 million tonnes of palm stems. It is necessary to conduct further research and technological developments to increase the added value of palm oil solid waste. Electricity generation must be decentralized in order to support rural electrification for the development of renewable energy (Wu et al. 2017). As most biomass energy comes from palm oil, which is approximately 98.7% of the total potential, relevant policies and innovative technologies are needed to utilize biomass effectively (Suzuki et al. 2017).

Biogas Energy

To mitigate emissions, one solution is to replace fossil fuel consumption with renewable energy resources. Various regulations and policies have been established for renewable energy development in order to accelerate this replacement. One of the potential developments is biogas that can be produced from various kinds of organic waste such as biomass, human and animal waste or from a mixture of plants such as oil palm (Purnamo, 2023). Indonesian Sustainable Palm Oil (ISPO) is a policy released by the Indonesia government to increase the competitiveness of Indonesian palm oil production and to increase the contribution to mitigate greenhouse gas emissions, leading to an environmentally friendly palm oil industry or sustainable palm oil industry (Pangarso and Endang, 2022). In palm oil mills, liquid waste, or POME, is a contaminative waste produced from three main processes: sterilization, condensation, and hydrocyclones. The high biological oxygen demand (BOD), chemical oxygen demand (COD), and acidity of POME can cause environmental problems if directly discharged into aquatic and terrestrial ecosystems without processing. The solution could be the initial process of POME and installation of anaerobic reactors for sustainable biogas production (Nasution et al. 2020). POME is the primary source of water pollution in the palm oil industry. Most palm oil mills are equipped with anaerobic and aerobic ponds to process POME in compliance with environmental regulations (Aziz et al. 2020). In recent years, palm oil mills have been equipped with biogas storage facilities to capture biogas and reduce greenhouse gas emissions. As biogas can be used as a biofuel, POME is considered a potential substrate for renewable energy sources. Biogas, which consists mostly of methane and carbon dioxide, is produced through the anaerobic digestion of POME. Biogas production involves the native microbial communities through hydrolysis, acidogenesis, acetogenesis, and methanogenesis (Lok et al. 2020). In addition to being used as a raw material for animal feed or fertilizer, POME can also be converted into biogas and used as a fuel to produce electricity. Biogas power plants that use POME as their raw energy material can provide economic benefits for palm oil producers not only by using it in internal processes but also by exporting their electricity into the grid (Rajani et al. 2019; Sari, 2019). Sumatera Utara Province, the fourth largest palm oil producer in Indonesia, has a large opportunity to use POME as a fuel for power plants.

A study (Hambali and Rivai, 2017) reported that in 2016, the Indonesia palm oil industry produced wastes of 31,380,742 tons of EFB, 17,891,231 tons of MF, 8,621,859 tons of PKS, 75,700,657 tons of POME, 128,914,621 tons of fronds, and 59,722,455 tons of trunks. In 2020, the palm oil industry produced wastes of 37,816,105 tons of EFB, 21,560,251 tons of MF, 10,389,975 tons of PKS, 91,224,865 tons of POME, 128,914,621 tons of fronds, and 59,722,455 tons of trunks. These wastes will continue to increase and in 2030, it is estimated that total production will be 53,904,512 tons of EFB, 30,732,801 tons of MF, 14,810,264 tons of PKS, 130,035,387 tons of POME, 128,914,621 tons of fronds, and 59,722,455 tons of trunks. The fronds and rods were similar in 2016, 2020, and 2030 because of a moratorium on expanding land use for palm oil plantations, as instructed by the President of the Republic of Indonesia in 2016.

Municipal Solid Waste Energy

Another potential renewable energy resource is municipal solid waste. In 2022, Sumatera Utara Province will be the 7th largest municipal solid waste-producing province in Indonesia (Aziz and Hanafiah, 2020). The total waste in the landfill is 53,802 m^3/day generated by a population of 15,372,437 people (Badan Statistik Sektoral, 2023). In general, the average physical composition of waste in the four landfills in Sumatera Utara Province consisted of 65% organic waste, 14% plastic waste, 8% paper waste, 3% LWTR, and 10% was classed as "other." The calorific values of the organic and inorganic wastes are 2464 and 4208 kcal/kg, respectively (Dalimunthe, 2018; Putri, 2018). Waste that can potentially produce electricity has 20-25% moisture by weight and a calorific value bigger than 2000 Kcal/kg. Technology for converting waste into electrical energy has been widely used in developed countries, such as Norway, Sweden, Germany, Switzerland, and England. Asian countries, including Japan, South Korea, Singapore, and Taiwan have also been developing this technology. These technologies include incineration, pyrolysis, gasification, and landfilling. The most commonly used technology is incineration, which is direct combustion using air as the oxidizer. Sumatera Utara Province has planned a regional waste management area based on the Presidential Regulation of the Republic of Indonesia No. 62 of 2011 concerning the administrative spatial planning of Sumatera Utara districts/cities, namely Medan, Binjai, Deliserdang, and Karo, known as MEBIDANGRO, which continues to be discussed with the public. This regulation provides an opportunity for the development of Waste Power Plant (PLTSa) installations to fulfill the government's plans, as stated in Presidential Regulation (Perpres) Number 35 of 2018 concerning the Acceleration of Development of Installations for Processing Waste into Electrical Energy Based on Environmentally Friendly Technology. The potential for municipal waste in the MEBIDANGRO area, which is capable of producing 121 MW of electric power per day, is shown in Table 1.

The development of the MEBIDANGRO waste management area is expected to provide a solution to urban waste problems, such as the Benowo waste power plant (PLTSa) in Surabaya, with a capacity of 9 MW and a waste requirement of 1000 tons/day, using gasification and landfill technology (Qodriyatun, 2021). In addition, there is the PLTSa *Merah Putih* in Bantargerbang, DKI Jakarta, which has an electrical energy capacity of 110.59 kWh/ton of waste. This PLTSa consumes 100 tons of waste per day and processes 100 tons of dry waste to produce 12 MW of electric power;

studies have been conducted to increase the electric power to 200 MW (Rajagukguk, 2020).

Table 1. The potency of municipal solid waste in MEBIDANGRO

Area/ Regency	kg/day (Wet condition)	kg/day (18% RDF)	kJ/kg	kJ/day	Joule/hr	Watt	MW
Medan	1.768.000	318.240	16000	5.091.840.000	212.160.000.000	58.933.333	59
Binjai	167.000	30.060	16000	480.960.000	20.040.000.000	5.566.667	6
Deli Serdang	1.078.000	194.040	16000	3.104.640.000	129.360.000.000	35.933.333	36
Karo	630.000	113.400	16000	1.814.400.000	75.600.000.000	21.000.000	21
Total	3.643.000	655.740	16000	10.491.840.000	437.160.000.000	121.433.333	121

Solar Energy

The other renewable energy resource that can be considered for exploration in Sumatra Utara Province is Solar Energy. The average global horizontal irradiation (GHI) in Sumatra Utara Province is 4,052 kWh/m^2 per day (Solargis, 2018). Global horizontal irradiation (GHI) is the total amount of shortwave irradiation received by horizontal surfaces on the ground. This value is important because it is used in the development and construction of solar power systems, such as photovoltaic solar panels, which use horizontal irradiation to convert electricity. GHI includes both normal direct irradiation (DNI) and diffuse horizontal irradiation (DHI). DNI is the radiation that comes directly from the sun, whereas DHI is the radiation scattered by molecules and particles in the atmosphere.

The Medan Meteorology, Climatology, and Geophysics Agency of Sumatra Utara province released data that were analyzed using the Hargreaves method. Data estimations were performed using the latitude and minimum and maximum monthly temperatures in Medan. The results show that the average monthly solar radiation is 6.47 kWh/m^2. This value is greater than the normal solar radiation of 3 kWh/m^2 under clear-sky conditions in Medan, and the radiation intensity is very high (Dang, 2017). This data shows that the solar energy in this area is capable of producing monthly electrical energy of 841.40 Wh/m^2 and if all land in Medan (265,100,000 m^2) is covered with horizontal PV panels, then almost 0.22 GWh of electricity can be generated per month. This reveals the significant potential of solar irradiation for PV power plants in Mediterranean (Saleh et al. 2017). Another technology that can be used is the Floating Photovoltaic (FPV) system, which is a PV system with several advantages, including saving land costs and hybrid connections with existing hydroelectric power plants. A simulation study using Google Erath and PVGIS software regarding the possibility of implementing an FPV system at the Sigura-Gura Dam, Sumatera Utara, with a total water surface area of 233,964 m^2 has been reported (Tarigan, 2022). The results showed that solar modules with a capacity of 12.6 MWp could be applied, with an annual energy production of 16.32 GWh.

Wind Energy

Wind is a popular source of renewable energy, not only as a standalone power generation but also in hybrid systems with other technologies. As a tropical country, Indonesia has a low average wind velocity, with an annual average of 3–6 m/s. A comprehensive analysis of the average wind speed in Indonesia is important for identifying locations suitable for offshore wind power plants. A study that employs spatial mapping approach to assess the average wind speed at a height of 100 m by combining global data from the European Center for Medium-Range Weather Forecasts Relysis v5 (ECMWF ERA5) has been found in literature (Poerbandono, 2016). The results for the period from 2013 to 2023 are presented in Figure 2. For the Sumatra Utara province case, there was no significant number of wind speeds in its offshore areas. The complexity of Indonesia's geographical conditions creates additional challenges in regard to determining the optimal locations for wind power plants. Therefore, each region requires different feasibility studies for wind power plants, particularly for large-scale plants (KESDM, 2015).

Table 2 shows wind velocity and power density in several potential areas in Indonesia. A study in seven districts of Sumatera Utara Province, namely Langkat, Serdang Bedagai, Labuhan Batu, Asahan, Central Tapanuli, Samosir and Karo in June - August 2017 carried out measuring the wind speed and temperature for 3 days in each location for 12 hours per day (Sembiring, 2018). The results reveal that the highest value of wind speed, specific power and total energy was Samosir Regency with an average wind speed of 5.89 m/s, an average specific power of 333.33 Watts and an average total energy of 14395 KWh. On the other hand, the district with the lowest wind speed, specific power and total energy values was the Karo Regency with an average wind speed value of 0.95 m/s, an average specific power of 1.68 Watts and an average total energy of 72.46 KWh. The total values of wind speed, specific power and total energy for the Sumatera Utara province were 3.34 m/s, 101.85 Watts and 4400.09 KWh, respectively. Based on this study, it can be concluded that the province of Sumatera Utara has a wind speed of 3.34 m/s. This is categorized as a fairly low wind speed. Thus, it is very difficult to produce energy on a large scale (Poerbandono, 2016).

Table 2. Wind velocity and power density at several locations in Indonesia (KESDM, 2015)

Resources potential	Wind Speed at 50m (m/s)	Wind Power density at 50 m (W/m^2)	Total Sites	Provinces
Poor	<3	<45	66	Maluku, Papua, Sumba, Mentawai, Bengkulu, Jambi, East Nusa Tenggara, West Nusa Tenggara, Southeast Sulawesi, North Sulawesi, and North Sumatra.
Low	3-4	<75	34	Central Java, Maluku, Yogyakarta, Lampung, South Kalimantan, East Nusa Tenggara, West Nusa Tenggara, South Sulawesi, North Sulawesi, Central Sulawesi, and North Sumatra.
Intermediate	4.1–5.0	75-150	34	Central Java, East Java, Yogyakarta, Bali, Bengkulu, East Nusa Tenggara, West Nusa Tenggara, South Sulawesi, and Central Sulawesi.
Good	>5	>150	19	Banten, Jakarta, Central Java, Yogyakarta, East Nusa Tenggara, West Nusa Tenggara, Southeast Sulawesi, North Sulawesi, and South Sulawesi.

Conclusion and Recommendations

Renewable energy resources and development play an important role in providing clean energy to replace fossil fuel energy. This study was conducted in order to increase the development of renewable energy in Sumatra Utara Province by collecting data and information from parties and reviewing the literature from previous research. The results of this study show that water resources have the greatest potential, followed by geothermal resources, biomass, and biogas. In the development of solar energy, implementation is still slow because investment is expensive. Furthermore, the development of the Minicipal solid waste will implement a regional waste management system called Medan, Binjai, Deli Serdang, and Karo (MEBIDANGRO), which is expected to solve the urban waste problem. Wind energy has been shown to have a weak potency because wind speeds throughout the Sumatra-Utara region are very low. To support the green technology program, the government has planned many development incentives by implementing renewable energy-based projects to encourage the economy from the energy sector. Based on this study, several recommendations are made to increase the development of renewable energy. First, we analyzed in depth the potential for renewable energy from urban waste, considering that raw materials are very abundant. Besides being able to produce electrical energy, it can also reduce piles of waste. Second, a hybrid technology can be developed using an on-grid system. Third, there are clear and measurable government policies and funding mechanisms to maintain and encourage renewable-energy investments.

Disclaimer

None.

References

Aditya, J. J. M. S. A. 2020. Studi Pemanfaatan Tenaga Listrik Di Pabrik Kelapa Sawit Pt. Prima Sauhur Lestari Pematang Kerasaan. *J. Teknol. ENERGI UDA*, vol. 09, no. 2, pp. 98–104, 2020.

Aziz, M., Kurniawan, T., Oda, T., Kashiwagi, T. 2017. Advanced power generation using biomass wastes from palm oil mills. *Appl. Therm. Eng.*, vol. 114, pp. 1378–1386. doi: 10.1016/j.applthermaleng.2016.11.031.

Aziz, M. M. A. 2020. Recent advances on palm oil mill effluent (POME) pretreatment and anaerobic reactor for sustainable biogas production. *Renew. Sustain. Energy Rev.*, vol. 119, no. November, p. 109603. doi: 10.1016/j.rser.2019.109603.

Aziz, N. I. H. A., Hanafiah, M. M. 2020. Life cycle analysis of biogas production from anaerobic digestion of palm oil mill effluent. *Renew. Energy*, vol. 145, pp. 847–857. doi: 10.1016/j.renene.2019.06.084.

Bayer, P., Rybach, L., Blum, P., Brauchler, R. 2013. Review on life cycle environmental effects of geothermal power generation. *Renew. Sustain. Energy Rev.*, vol. 26, pp. 446–463. doi: 10.1016/j.rser.2013.05.039.

Chyou, Y. P., Chang, D. M., Chen, P. C., Chien, H. Y., Wu, K. T., Chein, R. Y. 2020. Development of biomass gasification technology with fluidized-bed reactors for enhancing hydrogen generation: Part I, hydrodynamic characterization of dual fluidized-bed gasifiers. *Appl. Sci.*, vol. 10, no. 1. doi: 10.3390/app10010002.

Danish, Ulucak, R. 2020. Linking biomass energy and CO2 emissions in China using dynamic Autoregressive-Distributed Lag simulations. *J. Clean. Prod.*, vol. 250, p. 119533. doi: 10.1016/j.jclepro.2019.119533.

Dang, M. Q. 2017. Potential of solar energy in Indonesia. *Sol. Energy Indones.*, vol. 7, no. June, pp. 0–26.

Das, S., Lee, S. H., Kumar, P., Kim, K. H., Lee, S. S., Bhattacharya, S. S. 2019. Solid waste management: Scope and the challenge of sustainability. *J. Clean. Prod.*, vol. 228, pp. 658–678. doi: 10.1016/j.jclepro.2019.04.323.

Directorate of Food Crops, Horticulture, Indonesia Oil Palm Statistic 2022, *BPS - Stat. Indones.*, vol. 16, p. 32, 2023.

Dalimunthe, R. A. 2018. Studi Karakteristik Sampah Rumah Tangga di Kecamatan Medan Area dan Kecamatan Medan Polonia di Kota Medan (Timbulan, Karakteristik, dan Nilai Kalor). *Repos. Univ. Sumatra Utara*, 2018, [Online]. Available: https://repositori.usu.ac.id/handle/123456789/9206.

El Haj Assad, M., Ahmadi, M. H., Sadeghzadeh, M., Yassin, A., Issakhov, A. 2021. Renewable hybrid energy systems using geothermal energy: hybrid solar thermal–geothermal power plant. *Int. J. Low-Carbon Technol.*, vol. 16, no. 2, pp. 518–530. doi: 10.1093/ijlct/ctaa084.

Erivianto, D. 2018. Kajian Ekonomis Pemanfaatan Tandan Kosong Kelapa Sawit Sebagai Bahan Bakar Pltu Biomasa. *Semin. Nas. R.*, vol. 9986, no. September. [Online]. Available: https://jurnal.stmikroyal.ac.id/index.php/senar/article/view/212.

Fujita, H., NakanoO, T., Chaisomphob, K., Hambali, E. 2019. Greenhouse Gas Emission of Electricity Generation and Field Abandonment Scenarios Using Empty Fruit Bunches at a Palm Oil Mill, Surat Thani Province, Thailand. *J. Japan Inst. Energy*, vol. 98, no. 5, pp. 119–123. doi: 10.3775/jie.98.119.

Hambali, E., Rivai, M. 2017. The Potential of Palm Oil Waste Biomass in Indonesia in 2020 and 2030. *IOP Conf. Ser. Earth Environ. Sci.*, vol. 65, no. 1. doi: 10.1088/1755-1315/65/1/012050.

Indonesia Minister of Energy and Mineral Resources, 2007. Constitution No 30/2007. *Gov. Indones.* [Online]. Available: https://peraturan.bpk.go.id/Details/39928/uu-no-30-tahun-2007.

Indonesia Minister of Energy and Mineral Resources. 2016. Regulations of The President of the Republic of Indonesia No 4/2016. *J. Lemhannas RI*, pp. 1–40. [Online]. Available: http://download.garuda.kemdikbud.go.id/article.php?article=2961686&val=26383&title=Percepatan Pembangunan Infrastruktur Transportasi Laut.

Indonesia Minister of Energy and Mineral Resources. 2017. MEMR Regulation No. 12/2017. *Gov. Indones.*, no. 189, pp. 1–22.

Khwee, G. W. M. I. Y. K. H. 2018. Analisis Potensi Fiber Dan Cangkang Kelapa Sawit Sebagai Sumber Pembangkit Listrik Tenaga Uap (PLTU) Di PT. Pundi Lahan Khatulistiwa. vol. 74, no. 5, pp. 523–528.

Li, K., Liu, C., Jiang, S., Chen, Y. 2020. Review on hybrid geothermal and solar power systems. *J. Clean. Prod.*, vol. 250. doi: 10.1016/j.jclepro.2019.119481.

Lok, X., Chan, Y. J., Foo, D. C. Y. 2020. Simulation and optimisation of full-scale palm oil mill effluent (POME) treatment plant with biogas production. *J. Water Process Eng.*, vol. 38, no. July. doi: 10.1016/j.jwpe.2020.101558.

Mahmudah, N. 2018. Analisis Stabilitas Transient dengan Mempertimbangkan Respon Stabil dari Sudut Rotor, Frekuensi dan Tegangan di Sistem Sumatera Utara 150kV / 275kV Berdasarkan Rencana Usaha Penyediaan Tenaga Listrik (RUPTL) Tahun 2023. Departemen Teknik Elektro. ITS: Surabaya.

Minister of Energy and Mineral Resource. 2018a. MEMR Regulation No. 48/2018. *Gov. Indones.* [Online]. Available: https://peraturan.bpk.go.id/Details/142948/permen-esdm-no-48-tahun-2018.

Minister of Energy and Mineral Resources. 2018b. MEMR Regulation No. 50/2018. *Gov. Indones.* [Online]. Available: https://peraturan.bpk.go.id/Details/142950/permen-esdm-no-50-tahun-2018.

Minister of Energy and Mineral Resources. 2018c. MEMR Regulation No. 52/2018. *Gov. Indones.* [Online]. Available: https://peraturan.bpk.go.id/Details/142951/permen-esdm-no-52-tahun-2018.

Ministry of Energy and Mineral Resource. 2019. MEMR Regulation No. 12/2019. *Gov. Indones.* [Online]. Available: https://peraturan.bpk.go.id/Details/142275/permen-esdm-no-12-tahun-2019.

Ministry of Energy and Mineral Resource. 2019. MEMR Regulation No. 13/2019. *Gov. Indones.* [Online]. Available: https://peraturan.bpk.go.id/Details/142299/permen-esdm-no-13-tahun-2019.

Nasution, M. A., Herawan, T., Rivani, M. 2014. Analysis of Palm Biomass as Electricity from Palm Oil mills in North Sumatra. *Energy Procedia*, vol. 47, pp. 166–172. doi: 10.1016/j.egypro.2014.01.210.

Nasution, M. A., Wulandari, A., Ahamed, T., Noguchi, R. 2020. Alternative POME treatment technology in the implementation of roundtable on sustainable palm oil, Indonesian sustainable palm oil (ISPO), and Malaysian sustainable palm oil (MSPO) standards using LCA and AHP methods. *Sustain.*, vol. 12, no. 10. doi: 10.3390/su12104101.

Nazaruddin, N., Alkindi, A., Surdia, A. 2014. Analisa Kelayakan Pembangkit Listrik Tenaga Biomassa Sawit (PLTBS) PT. Perkebunan Nusantara I Aceh. *JURUTERA*. [Online]. Available: https://www.ejurnalunsam.id/index.php/jurutera/article/view/736.

Ompusunggu, B. G., Pamuji, M. R., Wibawa, G., Kuswandi. K. 2017. Studi Perencanaan Pembangunan Pembangkit Listrik Tenaga Panas Bumi Sebesar 120 MW di Blok Sarulla. *J. Tek. ITS*, vol. 5, no. 2, pp. 500–502. doi: 10.12962/j23373539.v5i2.16892.

Paes, M. X., de Medeiros, G. A., Mancini, S. D., Bortoleto, A. P., Puppim de Oliveira, J. A., Kulay, L. A. 2020. Municipal solid waste management: Integrated analysis of environmental and economic indicators based on life cycle assessment. *J. Clean. Prod.*, vol. 254, p. 119848. doi: 10.1016/j.jclepro.2019.119848.

Pangarso, S. S., Kusdiyantini, E. 2022. Potential Utilization of Biogas from Palm Oil Mill Effluent of PTPN 5: A Review. *18 Jmemme*. vol. 6, no. 01, pp. 18–31. doi: 10.31289/jmemme.v6i1.6298.

Parinduri, L. 2016. Analisa Pemanfaatan Biomassa Pabrik Kelapa Sawit Untuk Sumber Pembangkit Listrik. *JET (Journal Electr. Technol.*, vol. 1, no. 2, pp. 37–40.

Paulillo, A., Striolo, A., Lettieri, P. 2019. The environmental impacts and the carbon intensity of geothermal energy: A case study on the Hellisheiði plant. *Environ. Int.*, vol. 133, no. June, p. 105226. doi: 10.1016/j.envint.2019.105226.

Poerbandono. 2016. Wind characteristics and the associated risk of erosion in Seribu Islands patch reef complexes, Java Sea, Indonesia. in *AIP Conference Proceedings*. p. 080001, doi: 10.1063/1.4947416.

Pranoto, B., Pandin, M., Rahma, F. S., Nasution, S. 2013. Biomass Potential Map As a Database of National Scale Biomass Energy Development. *Ketenagalistrikan Dan Energi Terbarukan*, vol. 12, no. 2, pp. 123–130.

President of Republik Indonesia. 2022. Regulations of The President of the Republic of Indonesia No 112/2022. *Gov. Indones.* [Online]. Available: https://peraturan.bpk.go.id/Details/225308/perpres-no-112-tahun-2022.

Purnomo, T. 2023. Overview of biogas development as alternative energy in Indonesia. in *Proceedings, AIP Conference*. p. 030053, doi: 10.1063/5.0124862.

Putra, A. D., Yusuf, I., Gani, U. A. 2016. Studi Potensi Limbah Biomassa Kelapa Sawit Sebagai Pembangkit Listrik Tenaga Uap (PLTU) Di PT. Perkebunan Nusantara Xiii PKS Parindu. *Univ. Tanjungpura Pontianak.*

Putri, C. N. 2018. Studi Karaktersitik Sampah Rumah Tangga di Kecamatan Medan Helvetia dan Kecamatan Medan Selayang di Kota Medan (Timbulan, Komposisi, Karakteristik dan Nilai Kalor). *Repos. Univ. Sumatra Utara*. [Online]. Available: https://repositori.usu.ac.id/handle/123456789/9218.

Qodriyatun, S. N. 2021. Pembangkit Listrik Tenaga Sampah: Antara Permasalahan Lingkungan dan Percepatan Pembangunan Energi Terbarukan. *Aspir. J. Masal. Sos.*, vol. 12, no. 1, pp. 63–84. doi: 10.46807/aspirasi.v12i1.2093.

Rajagukguk. J. R. 2020. Studi Kelayakan Desain Pembangkit Listrik Tenaga Sampah (PLTSa) Sebagai Sumber Energi Listrik 200 MW. *Media Ilm. Tek. Lingkung.*, vol. 5, no. 1, pp. 51–61. doi: 10.33084/mitl.v5i1.1371.

Rajani, A., Kusnadi, Santosa, A., Saepudin, A., Gobikrishnan, S., Andriani, D. 2019. Review on biogas from palm oil mill effluent (POME): Challenges and opportunities in Indonesia. *IOP Conf. Ser. Earth Environ. Sci.*, vol. 293, no. 1. doi: 10.1088/1755-1315/293/1/012004.

Renstra KESDM 2015-2019. *Badan Penelit. dan Pengemb. ESDM*, 2015.

Republic of Indonesia. 2014. Peraturan Pemerintah Republik Indonesia No.79 Tahun 2014 Tentang Kebijakan Energi Nasional. *Huk. Online*, pp. 1–60.

Safrizal. 2015. Small Renewable Energy Biogas Limbah Cair (Pome) Pabrik Kelapa Sawit Menggunakan Tipe Covered Lagoon Solusi Alternatif Defisit Listrik Provinsi Riau. *J. DISPROTEK*, vol. 6, no. 1, pp. 26–35.

Sakidiansyah. 2011. Evaluasi Kinerja Pembangkit Listrik Tenaga Mikro Hidro di Desa Buluh Awar Kecamatan Sibolangit Kabupaten Deli Serdang Propinsi Sumatera

Utara. *https://repositori.usu.ac.id*. [Online]. Available: https://repositori.usu.ac.id/handle/123456789/79592.

Saleh, K., Irwanto, M., Hamid, H. A., Ismail, B., Nair, G., Alam, H., Masri, M., Yusoff, M. I. 2017. Estimation of Solar Irradiation in Medan Using Hargreaves Method Based on Minimum and Maximum Temperature for Potential Assessment of Photovoltaic Power Generation. *Adv. Sci. Lett.*, vol. 23, no. 5, pp. 4463–4466. doi: 10.1166/asl.
2017.8854.

Sari, S. P. 2019. Optimising the utilisation of Palm Oil Mill Effluent (POME) for biogas power plants to achieve Indonesian target of bioenergy power plants in 2025. *Univ. Twente*, no. August. [Online]. Available: http://essay.utwente.nl/79483/.

Sembiring, A. 2018. Pemetaan Potensi Energi Angin di Sumatera Utara. 2018.

Statistik Sektoral. 2023. *Dinas Komunikasi Dan Informatika Provinsi Sumatra Utara, Medan, 2023.* [Online]. Available: https://sumutprov.go.id/content/userfiles/statistiksektorial/Buku_Statistik_Sektoral_2023.pdf.

Stocks, M., Blakers, A., Cheng, C., Lu, B. 2019. Towards 100% renewable electricity for Indonesia: The role for solar and pumped hydro storage. *2019 Int. Conf. Technol. Policies Electr. Power Energy, TPEPE 2019*. doi: 10.1109/IEEECONF48524.2019.9102581.

Susanto, H., Suria, T., Pranolo, S. H. 2018. Economic analysis of biomass gasification for generating electricity in rural areas in Indonesia. *IOP Conf. Ser. Mater. Sci. Eng.*, vol. 334, no. 1. doi: 10.1088/1757-899X/334/1/012012.

Suzuki, K., Tsuji, N., Shirai, Y., Hassan, M.A., Osaki, M. 2017. Evaluation of biomass energy potential towards achieving sustainability in biomass energy utilization in Sabah, Malaysia. *Biomass and Bioenergy*, vol. 97, pp. 149–154. doi: 10.1016/j.biombioe.2016.12.023.

Tarigan, E. 2022. Simulation of the Floating PV System at Sigura-Gura Hydropower Dam in North Sumatra, Indonesia. in *2022 11th Electrical Power, Electronics, Communications, Controls and Informatics Seminar (EECCIS)*. Aug. pp. 71–75. doi: 10.1109/EECCIS54468.2022.9902912.

Wu, Q., Qiang, T. C., Zeng, G., Zhang, H., Huang, Y., Wang, Y. 2017. Sustainable and renewable energy from biomass wastes in palm oil industry: A case study in Malaysia. *Int. J. Hydrogen Energy*, vol. 42, no. 37, pp. 23871–23877. doi: 10.1016/j.ijhydene.2017.03.147.

Chapter 9

Developments in a Practical Green Economy in North Sumatra

Zulkifli Nasution*
and Marheni
Department of Agrotechnology, Faculty of Agriculture Universitas, Sumatera Utara, Medan, Indonesia

Abstract

A green economy has not been widely practiced in Indonesia, especially in North Sumatra. However, this issue is increasing as people find it easier to learn on social media. The green economy is known at the provincial level for preserving forest resources, and the Global Risk Perception Survey indicates that five of the ten most significant risks in the next 10 years will originate from environmental risks. Low carbon and climate-resilient development is "an obligation" in Indonesia's economic transformation toward a green economy. A green economy is part of an economic transformation strategy to push Indonesia out of the "middle-income trap" before 2045. Many national-level companies donate funds through corporate social responsibility to replant damaged forests; additionally, many foreign companies or institutions collaborate with national and local non-governmental organizations to rehabilitate mangroves and forests outside protected forests controlled by the state, contributing to the green economy. Such initiatives are aimed at changing the business work system that has been implemented thus far into a more innovative, collaborative, and sustainable system. Implementing a green economy is only possible if we consider implementing a system that will include the community, especially rural

* Corresponding Author's Email: zulnasution@usu.ac.id.

In: Natural Resources, Environment, and Food and Energy Security …
Editors: Mohammad Basyuni and Severino Salmo III
ISBN: 979-8-89530-283-5

communities, to meet as many of their basic needs as possible. Therefore, with the aim of a green economy, efforts are being made to enable the community to obtain new jobs and ensure that they do not use wood and forest resources as a livelihood to meet their daily living needs.

Keywords: green economy, environment, carbon, North Sumatra

Introduction

In principle, a green economy is characterized by low carbon emissions, resource efficiency, and social inclusivity. Economic growth is driven by government and private investments that reduce emissions, increase energy use efficiency, and prevent the loss of natural resources (Rudiyanto, 2022). In North Sumatra, this is implemented through various initiatives, such as planting damaged mangrove forests, establishing community forests, and promoting agroforestry (UNEP, 2024). The principles of a green economy based on a model centered on human activities and nature can be summarized as follows: well-being, justice, good governance, and poverty eradication (D'Amato et al. 2017).

A green economy is crucial for improving community welfare in a societal model. This goes beyond providing income for living needs and enhancing social and cultural quality. The preservation of the environment, a vital aspect of a green economy, is a culture of the community that must continue to be explored and implemented. This should be included in the formal education of the local curriculum. For example, the people of Mandailing Natal Regency in North Sumatra preserve fish by creating a model of a forbidden pit that can only be harvested on the third day of the Idul Fithri holiday. A relatively hefty fine is imposed if fish are captured outside of this schedule. The fine is used as cash or village financing to ensure sustainable river preservation. This approach helps protect the environment and fosters a sense of community and shared responsibility, making individuals empathetic and supportive of such initiatives.

A green economy encourages good organizational governance as it is based on accountable and transparent institutions. This will prevent arbitrariness in decision-making as a green economy requires community support and agreement; this will create collaboration between leaders or rule-makers and the community. For example, in the case of planting mangrove forests, there must be a community group of seeders, planters, and individuals who care for the plants. Thus, an agreement must be reached regarding the extent to which they will receive wages and other funding. For how long should this be maintained? Such commitments are conducted in writing, as required by donors who provide funds for planting or conservation. This encourages delegated decision-making and open dialogue to ensure that personal interests are not involved in the decision-making process. Such broad community support promotes a collaborative approach to problem-solving.

Adaptation to a green economy will require new skills or improvement of existing skills to be more environmentally friendly; therefore, new training and reformulation of school and college curricula are required to produce graduates who better understand the concepts and application of science within the framework of sustainable development. With more massive environmental campaigns and increasingly stringent regulations regarding the environment, industries becoming increasingly aware of the importance of environmental preservation have created new jobs in terms of ecological management and monitoring. Energy efficiency, as a concept of environmental management and reducing, reusing, and processing waste produced by industries and households, has encouraged new businesses in the green economy. Government regulations on ecological management requiring technical standards and environmental impact studies have indirectly increased the number of people involved in green economies.

In low-carbon development, renewable energy sources are used in a green economy wherever possible, such as solar energy, hydroelectric power, and hydrogen, which produce minimal CO_2 emissions. The energy use for water is considerably higher, particularly in power plants in North Sumatra. Power plants at levels I to IV in Sigura-gura Sungai Asahan have a large capacity and are a source of industrial electricity for PT Indonesia Asahan Aluminum in Kuala Tanjung, Batu Bara Regency. A large-capacity power plant is being constructed on the Batang Toru River in the South Tapanuli Regency. Hydroelectric power plants with small-to-medium capacities have been built in the North Tapanuli Regency, Central Tapanuli, and other areas. Processing waste into compost has been widely conducted by environmental services, the community, and plantation companies, which turn empty palm oil plantations into good-quality compost.

The green economy program in North Sumatra holds significant promise for the local population. Promoting economic growth that is both environmentally sustainable and resource-efficient can help reduce poverty and improve the quality of life. Transitioning to a green economy can create new job opportunities and stimulate economic growth, particularly by developing renewable energy sources and promoting sustainable industries. This benefits both the environment and the local economy, making the transition to a green economy a win-win situation for North Sumatra.

Low-carbon development: a green economy is based on the use of renewable energy sources, such as solar, wind, hydroelectric, and hydrogen, which generate little or no amounts of CO_2 emissions. A low-carbon economy encourages extensive electrification, with energy being produced sustainably from renewable sources in industry, mobility, and other aspects of life. The Green Growth Program supports Indonesia in realizing green economic growth, which can reduce poverty and ensure social inclusion, environmental sustainability, and resource efficiency (Bappenas, 2013).

The challenge of realizing a green economy and blue skies through low-carbon development includes the formulation of funding policies from public funds and private investment, including the development of the compensation the industry will receive if it implements a low-carbon program. Accelerating technology transfer and innovation in low-carbon industrial processes and developing affordable technologies must be carefully designed. The energy transition is a matter of engineering and human resources in line with technological changes (Rudiyanto, 2022).

Method

Flow Diagram

Flow diagrams are necessary in research and have a very important function to ensure that the research process runs in a systematic and organized manner. Figure 1 shows the flow diagram method used in writing this chapter.

Research Design

This type of research is a literature review. The data source used in this chapter was secondary data obtained from previous articles, journals, or internet sites. A literature review was conducted based only on written work, including the results of sound research and whether or not it had been published (Mestika, 2008).

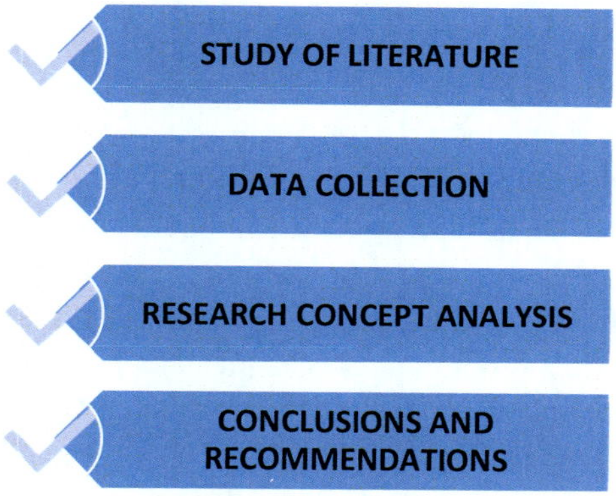

Figure 1. Flow diagram of this chapter.

Data Collection

The data obtained in this chapter were derived from e-books, journals, scientific articles, literature reviews, official government publications, or international institutions in the form of internet e-books obtained using Google scholar and Google searches.

Data Analysis

Data analysis was conducted based on individuals with more than five years of experience in managing coastal conservation, planting mangroves, fostering environmental awareness in rural communities, and sustainable development.

Results and Discussion

Indonesia Green Economy

The Ministry of National Development Planning (Bappenas) has planned a program for a green economy by developing projects that include banking based on national contributions that are adjusted to achieve sustainable development goals. Thus, the government links low-carbon projects with existing funding sources through statutory regulations to ensure that investment can be implemented; this guarantees green economic growth in sectoral planning at the district, provincial, and national levels. Thus, the government has designed innovative financial instruments and policies to transfer funds to sectors with shallow risk.

Based on the analysis of Bappenas, at least 6% of economic growth is required to ensure that Indonesia fulfills its vision of 2045. This economic transformation strategy with the green economy is a "game changer" that aims to create advanced and inclusive economic growth while realizing community welfare and maintaining quality. Low-carbon development and climate resilience policies are designed as the main instrument or "backbone" in implementing the green economy strategy. Studies have observed that if the green economy program is implemented, almost 100 billion tons of CO_2-eq will be saved in 2021–2060, and the emission intensity will reduce by up to 68% in 2045 before reaching zero by 2060. Additionally, this program is estimated to open 1.8 million environmentally friendly job opportunities (Putri et al. 2022).

As a country committed to controlling the rate of climate change, Indonesia must follow Article 3.4, namely, parties have the right to encourage sustainable development. Policies and measures to protect against climate change caused by human activities must be adapted to each region's conditions and integrated into national development programs. This must be implemented by considering that economic development is significant; therefore, measures can be adopted quickly to address climate change (United Nations, 1992).

Indonesia has submitted an updated National Contribution Document (NDC) and a long-term low-carbon and climate resilience strategy document for 2050. To achieve the 2030 NDC target and accelerate transformational change (KLHK, 2021) toward low-carbon development and climate resilience, Indonesia developed an implementation strategy for NDC comprising the following nine programs:

1. Establishing commitment between ministries and other government institutions, including regional government, the private sector, civil society, and financial institutions (stakeholders).
2. Increasing the capacity of institutions and human resources at all levels, encompassing all aspects related to climate change, especially mitigation and adaptation actions, and funding to halt the rate of climate change.
3. Creating a situation with more stakeholder involvement in mitigation, adaptation, and resource mobilization by implementing appropriate regulations, policies, and actions.
4. Developing a coordination network and establishing synergy between sectors, regions, and stakeholders.
5. Creating a single policy for greenhouse gases and implementing it transparently.
6. Developing intervention policies for NDC implementation, programs to integrate mitigation into five sectors (forestry, energy, waste, and agriculture), and adaptation systems both sectorally and regionally in development planning to obtain financial support from public funds and facilitate the mobilization of domestic and international resources.
7. Developing NDC implementation guidelines in ministries and other government institutions, both local governments and stakeholders, to ensure that they can implement program no. 6.
8. NDC implementation refers to the implementation of policies, plans, and intervention programs (program no. 6), considering the guidelines developed in program no. 7.
9. Monitoring the progress of NDC implementation, making adjustments to the NDC if necessary, and the formation of the monitors by the ministry.

Green Economy in North Sumatra Province

The green economy in North Sumatra Province is still improving; however, the actions required for its improvement are still uncertain. Our understanding of a green economy is limited to understanding and implementing forest conservation in the form of agroforestry and mangrove planting at suitable locations.

In reality, such conservation measures are only implemented by universities in the form of community service, community groups, and local industries. Such measures are conducted on a relatively large scale, by international institutions, in collaboration with regional and national NGOs. For example, one national NGO planted 2,435 ha of mangroves in North Sumatra and Aceh, equivalent to 7,887,680 seedlings, with a success rate of approximately 75%, which is an increase from the minimum success rate of 60% in 2 years (Yakopi, 2023).

No success will be achieved in efforts to conserve natural resources, such as mangroves, and safeguard biodiversity without the identification of replacement financial resources for the subsistence needs of rural communities. Rural communities initially obtained such resources by cutting down mangrove wood and encroaching on forests, both primary and secondary. The community must be educated to seek income sources from the non-timber forestry sector, such as harvesting benzo resin (*Styrax paralleloneurum*) or making sugar from nipa trees (*Nypa fruticans*). Nature conservation occurs if the community can obtain an adequate income without destroying nature. This is why a green economy is vital. For example, in Langkat Regency, an NGO has trained the community to make palm sugar and finds ways to market it if the community cannot do so. In the Deli Seran area, some NGOs have educated the community to develop a cracker industry, with permission from the Food and Drug Monitoring Agency, and have marketed it at relatively reasonable prices. Without a replacement source of income, the community will return to encroaching on the forest as a rapid source of income. Forest resources are also used for survival. Individuals from rural communities primarily destroy nature to meet their daily needs and not to get rich, such as businesspeople.

The importance of the agricultural sector in North Sumatra's economy is the focus of the government's efforts to spur economic growth. A green economy is based on improving green investment, sustainable management of assets and infrastructure, ensuring fair and affordable transformation, and empowering human resources. According to the United Nations Environment Program, "the green economy" prioritizes equality and development. Consequently, a green economy can be seen as a way to achieve environmental and resource justice, as the well-being of all societies should be the ultimate goal of all business endeavors (Nurul and Harahap, 2023).

Unfortunately, the development of a green economy may not always occur smoothly owing to the difficulties of bureaucracy, which does not yet understand its benefits, and old habits that always complicate work in the field. A green economy is defined as a low-carbon, resource-efficient, and socially inclusive economy. In a green economy, the growth in employment and income is driven by public and private investments in economic activities, infrastructure, and assets that reduce carbon emissions and pollution, enhance energy and resource efficiency, and prevent the loss of biodiversity and ecosystem services (UNEP, 2024).

This idea of saving the environment from human greed is beautiful. However, even at the government level, the concept of the green economy is not yet fully mature, as indicated by the lack of adequate laws and regulations at the central government and central regional levels.

From a bureaucratic perspective, the government has not yet understood the green economy, and state officials still view all community and NGO activities as habitual. The implementation of environmental protection, which is closely related to a green economy, is still difficult, even though it does not use government funds.

Instead of making implementation more accessible, various challenges and high costs are often experienced. In addition to government-related challenges, people are used to old and non-beneficial habits, which result in high costs for activities that should not occur. Until the government's attitude changes, the implementation of a green economy will progress slowly. Thus, we need to improve our behavior toward a better, non-corrupt, and respectable environment in all aspects of life.

The green economy incorporates a long-term view of the economy, emphasizing how resources are used to benefit society and how efficiency and resource sufficiency can be leveraged to foster wealth, resilience, and well-being for present and future generations while staying within the ecological bounds of the planet. A green economy exhibits inclusive and non-discriminatory behaviors, ranging from environmental activities to investments for well-being. Environmentalists cited this economic theory as the only means of preserving humanity. When the globe is colored green, it is more sustainable, inclusive, and capable of fending against the obstacle of global warming.

Conclusion

The implementation of a green economy in North Sumatra, Indonesia, is an important step in meeting the challenges of climate change and environmental risks. Although the green economy has not been fully implemented, public awareness is increasing, especially through social media. Green economy development in Indonesia aims not only to reduce carbon emissions, but also to improve resource use efficiency and create social equity. Initiatives such as mangrove forest rehabilitation and community forest planting demonstrate a commitment to integrating sustainability into economic practices. The importance of collaboration between the government, private sector and civil society was also emphasized, where all parties must be involved in mitigation and adaptation efforts to climate change. With the right strategy and support from various stakeholders, Indonesia can achieve the set national contribution targets and transform towards a more sustainable and inclusive economy.

Disclaimer

None.

References

Bappenas. 2013. "*Indonesia Green Growth Program.*" http://greengrowth.bappenas.go.id/tentang-kami/. Accessed on: 16th April 2024.

D'Amato, D., Droste, N., Allen, B., Kettunen, M., Lahtinen, K., Korhonen, J., Leskinan, P., Matthies, Brent. D., Toppinan, A. 2017. Green, Circular, bio economy: a comparative analysis of sustainability concepts. *Journal of cleaner Production* 168(1): 1-31.

Director General for Climate Change as Indonesian NFP to the UNFCCC. 2021. "*Updated Nationally Determined Contribution Republic of Indonesia Ministry of Environment and Forestry Directorate General of Climate Change.*"

Falatehan, A. F., Syaukat, Y., Hastuti, H., Nasrullah, N. 2021. Paddy Loss and Its Implication to Fertilizer Subsidy in Indonesia. HAYATI *Journal of Biosciences* 28(1): 73–73.

Mestika, Z. 2008. *Metoda Penelitian Kepustakaan.* Yayasan Pustaka Obor Indonesia.

Nurul, A., Harahap, I. 2023. Analisis Pengembangan Ekonomi Hijau Dengan Basis Pertanian Dengan Implementasi Maqashid Syariah Di Sumatera Utara" (Analysis of Green Economic Development Based on Agriculture with the Implementation of

Maqashid Syariah in North Sumatra). *Al-Kharaj: Jurnal Ekonomi, Keuangan & Bisnis Syariah* 5(5): 2535-2543 P-ISSN 2656-2871 E-ISSN 2656-4351. DOI: 10.47467/alkharaj.v5i5.3505

Putri, L. A., Napitupulu, A., Amalia, A., Putri, A. P., Armanto, A. N., Ramdani, D. A., Suarga, E., Damayanti, H., Darliazi, I., Medrilzam, Nasution, M. I., Gardian, P. I., Firmansyah, R., Ida, S., Dwitiyasih, T. 2022. "Green economy index a step forward to measure the progress of low carbon & green economy in Indonesia." *The Ministry of National Development Planning (Bappenas)*. Ekawati, I. 2010. Pengembangan Potensi Jagung Menuju Revitalisasi Pertanian di Kabupaten Sumenep. *Jurnal Pertanian CEMARA*, 7(1): 25–33.

Rudiyanto, A. 2022. "Green Economy: Arah Baru Transformasi Ekonomi Indonesia yang Berkelanjutan" (Green Economy: New Direction for Sustainable Indonesian Economic Transformation). *Presented at the Bank Indonesia Main Leadership School Program (SES TABI) VII 2022: "Building Transformative Leadership Towards a Leading Digital Central Bank to Accelerate Inclusive and Sustainable Development,"* Ministry of National Development Planning, Bappenas, Indonesia.

UNEP. 2024. "*Green Economy.*" https://www.unep.org/regions/asia-and-pacific/regional-initiatives/supporting-resource-efficiency/green-economy. Accessed on: 17th April 2024

UNEP. 2024. "*Green Economy*" https://www.unep.org/regions/asia-and-pacific/regional-initiatives/supporting-resource-efficiency/green-economy. Accessed on: 16th April 2024.

United Nations. 1992. "*United Nations Framework Convention on Climate Change.*"

Yakopi. 2023. *Indonesian Conservation Foundation Report*. Not published.

Chapter 10

The Contribution of Oil Palm Plantations to Prosperity and a Circular Economy in North Sumatra

T. Sabrina[1,3,*]
and Hotnida Sinaga[2]

[1]Department of Natural Resources and Environment, Postgraduate School, Universitas Sumatera Utara, Medan, Indonesia
[2]Department of Food Technology, Faculty of Agriculture, Universitas Sumatera Utara, Medan, Indonesia
[3]Department of Agrotechnology, Faculty of Agriculture, Universitas Sumatera Utara, Medan, Indonesia

Abstract

Oil palm plantations have great potential for economic development, both at the local and national level, especially in the North Sumatra region. The development of the palm oil plantation industry has a positive impact on the food sector, increasing economic income and creating jobs, as well as developing infrastructure on a wider scale. The content of palm oil plays an important role in the making of various product food like bettering the surface and stability of food products as well as making food safer from oxidation damage. The increase in farmers' income since the shipment of palm oil exports has contributed greatly to strengthen financial solidity and supporting the improvement of development in Indonesia. In its expansion, the palm oil industry directly provides millions of jobs, both in plantations and industrial facilities, and also has implications for related sectors such as

* Corresponding Author's Email: t.sabrina@usu.ac.id.

In: Natural Resources, Environment, and Food and Energy Security …
Editors: Mohammad Basyuni and Severino Salmo III
ISBN: 979-8-89530-283-5

transportation, trade and logistics. However, there are also challenges and negative impacts that need to be managed, such as natural impacts, conflict issues between plantation companies and surrounding communities, and social welfare that is still not enjoyed by workers/farmers. In terms of natural impacts, palm oil plantations are often criticized for causing deforestation, loss of biodiversity, and increase poverty. With regard to social welfare, despite creating jobs, working conditions in plantations often come under scrutiny, including compensation issues as well as poor working conditions and violations of workers' human rights.

Keywords: palm oil, prosperity, environmental impact, plantations

Introduction

It is an amazing sight, for most people working in the early stages of opening a palm oil plantation, to see the transformation from an uninhabited area to a thriving village and then to a town, and this is due to the growth of the palm oil industry. There are many workers who are employed to support the industry, although the number is not as many as that in the seasonal plantations such as tobacco and sugarcane. This in turn has led to the development of other jobs, shops, banks and markets, which are necessary to meet the needs of these workers. Palm oil (*Elaeis guineensis*) is an important food and commodity crop cultivated in many humid tropical regions (Byerlee et al. 2017). Although the plant is not native to Southeast Asia, there has been a substantial increase in palm oil expansion in the region, particularly in Indonesia and Malaysia, which are the world's leading producers and exporters of crude palm oil (Qaim et al. 2020).

Several studies have shown that palm oil production has contributed to significant income increases for smallholder households (Ayompe et al. 2021), while stimulating rural and community development (Santika et al. 2019). This is what makes the development of oil palm plantation areas in Indonesia very fast, even changing the function of land from forest land, paddy fields, or other agricultural crops to palm oil plantations.

The growth of oil palm land area in Indonesia continues to increase, where the area growth rate during 2004-2014 was 7.67%, or around 10.9 million ha (Directorate general of plantation, 2014). According to data on national superior plantation statistics in 2020-2022, the palm oil area reached 15 million ha. The increase occurred by 5 million ha within 8 years. Palm oil area is dominated by private plantations, followed by smallholder plantations and national plantations. The increase in area is found in private plantations by millions of ha and smallholder plantations by hundreds of thousands of ha, while national plantations increase by only tens of thousands of ha (Directorate general of plantation, 2016).

The ten centers of palm oil producers in Indonesia are Riau, West Kalimantan, North Sumatra, Central Kalimantan, East Kalimantan, Jambi, South Sumatra, Aceh, South Kalimantan, and West Sumatra. Interest in developing oil palm land is certainly accompanied by the promise of economic growth and how much rupiah will be obtained. Although North Sumatra province is ranked the third in terms of oil palm land area, this province is the first area to open oil palm plantations. In 1911, Hallet built a 2630 ha oil palm plantation in the Eastern Sumatra region including Pulu Raja (Asahan district) and Sungai Liput (Aceh Tamiang) (Indonesian oil palm research institute, 2016).

Crude palm oil (CPO) production in Indonesia in 2022 amounted to 48,235,405 tons, dominated by private plantations followed by smallholder plantations, and the least contribution was national plantations. When compared to the production of other plantation crops, palm oil production is the highest in Indonesia. This proves that the palm oil industry plays an important role in Indonesia's agriculture and economic growth. As a prime crop, the industry contributes significantly to gross domestic product, gross national income, foreign exchange and employment.

Indonesia exports CPO and its derivatives to more than 45 countries worldwide. The nine largest consumers of Indonesian CPO are India, the European Union, China, Malaysia, Singapore, Bangladesh, Egypt, Pakistan and the United States. In 2009, India, China and the EU accounted for 52% of global imports. As much as 80% of palm oil production comes from developing countries, so it is an important factor in increasing people's income. World palm oil production is currently dominated by Indonesia and Malaysia, accounting for around 85-90% of total production. Projections of global vegetable oil demand in 2024 are increasing for food needs, the oleochemical industry, and energy needs with the implementation of B35 biodiesel on a full-year basis. Therefore, it is estimated that global vegetable oil demand will reach more than 307 million tons, an opportunity for Indonesian CPO (Sindonews.com).

According to Asian Agri organization, the palm oil industry in Indonesia employs more than six million workers, more than half of whom are directly involved in plantations. The palm oil sector in Indonesia attracts 5,000 to

6,000 new workers each year and requires educated and skilled human resources in its processing. Although the industry has been around for decades before Indonesia's independence, the palm oil industry still remains a major employer in Indonesia. Moreover, sustainability and the environment are also receiving increasing attention in the industry.

In line with the development of palm oil, new towns grew up around the plantation areas, in which there are new economic activities in the new area. On the other hand, the development of palm oil plantation coconut correlates with a decrease in forest land area and will certainly pose a threat to the environment, such as a decrease in the biodiversity of flora and fauna in the area. Prosperity is generally translated as a condition of successful development, especially in relation to economic welfare. A broader understanding of prosperity must be linked to the environment. This review article discusses the contribution of the palm oil industry in realizing prosperity and economic turnover in North Sumatra.

The Palm Oil Industry in Indonesia

The dynamics of Indonesia's palm oil exports depend on the strength and weakness of global demand. In November 2017, crude palm oil (CPO) exports from the world's top producer Indonesia declined due to reduced production and weak demand from key markets China and India. CPO production fell to 3.90 million tons in November 2017 from 3.95 million tons in November 2017. October 2017 was the highest monthly level achieved since the survey began in 2014. CPO exports in November 2017 are estimated to decline to 2.52 million tons from 2.75 million tons in October 2017. India's increase in palm oil import tax reduced the country's imports in November 2017 to the lowest level in eight months. Chinese demand is also expected to ease into 2018 ahead of the Lunar New Year. European and other regional markets also show weak demand. Indonesia's palm oil and palm kernel oil exports totaled 2.60 million tons in October 2017, according to data from the Indonesian Palm Oil Association (GAPKI). The country's local consumption was reported at 989,550 tons in November 2017 compared to 985,000 tons in October 2017. The decline in exports pushed stockpiles to 3.60 million tons in November 2017 from 3.14 million tons (Original source: the star, 2017).

Fluctuations in exports of CPO do not really interfere with palm oil production. This is because the domestic demand for palm oil is quite high.

Palm oil production has grown to meet the increasing demand for vegetable oil. Palm oil production has increased rapidly since the 1960s. Between 1970 and 2020, global palm oil production increased about 40 times. Global production increased from just 2 million tons to around 80 million tons, with Indonesia producing more than half of this. A

Oil palm crop is produced in a number of countries including Thailand, Colombia and Nigeria. As is expected, all of these countries are located along the zone of 'optimal conditions' around the equator. Palm oil is grown in small quantities in many countries, but the global market is dominated by only two countries: Indonesia and Malaysia.

Why is the market for palm oil and vegetable oils increasing so rapidly? What is it used for? Palm oil is extremely versatile and is used in a variety of products around the world in food, industrial applications and bioenergy. From shampoos to detergents, and from chocolate to cookies, many companies use palm oil in a variety of branded products. More than two-thirds (68%) is used in foods ranging from margarine to chocolate, pizza, bread, cooking oil and animal feed. Retailers of fried tofu, bananas and other micro, small and medium enterprises (MSMEs) depend on the availability of palm oil. In fact, palm oil has recently become a substitute fuel for petroleum, with as much as 5% used as biofuel for transportation, electricity, or heat. While food products dominate globally, the breakdown varies by country. Some countries use more palm oil for biofuels than others. In Germany, for example, bioenergy is the largest use, at 41% (more than food at 40%). The push towards increased biofuel consumption in the transportation sector has encouraged this, despite its worse environmental impact than typical diesel fuel.

Plantations have a large role and potential in Indonesia's economic development. Agricultural plantation development policies are focused on plantation commodities that play an economic, social and environmental role. Oil palm plants are included in the top priority commodities. The role of competent human resources in managing good plantation management requires good institutional governance which requires the support of competent human resources (Annual report BPDPKS, 2021).

North Sumatra is included as a plantation in North Sumatra has been opened since the Dutch colonial period. Some of North Sumatra's leading plantation commodities include rubber, oil palm, coffee, coconut, cocoa, tobacco, and sugarcane. In 2020, the Central Bureau of Statistics (BPS) recorded the largest area of smallholder oil palm plantations in North

Sumatra, namely 440 thousand ha with production reaching 7 million tons (Central statistics agency, 2020).

Palm Oil in Food and Nutrients

About 80% of palm oil, palm kernel oil and their fractions produced globally are used for food purposes. Palm oil can be categorized into Crude Palm Oil and Palm Kernel Oil with significantly different properties. Crude Palm Oil is crude palm oil obtained from the extraction of palm fruit pulp. The pulp is separated from the seeds during the extraction process, and the separated seeds are further processed to produce oil (Quality Indonesia, 2022). Palm Kernel Oil is the oil produced from the palm kernel. When compared to other palm oils, the texture of Palm Kernel Oil is thicker and denser (Hasibuan, 2021). Palm oil is the best choice for cooking and baking fats such as margarine, butter, and vanaspati. Palm kernel oil, which is derived from the flesh of the palm kernel, has a high lauric acid content and has a sharp melting point, making it suitable for use in confectionery fats. Fractionation of palm oil into palm stearin and palm olein, and palm kernel oil into palm kernel stearin and palm kernel olein further increases the use of palm oil and palm kernel oil in food. Palm oil and palm oil olein, due to their high content of oleic acid and natural antioxidants (vitamin E), have excellent oxidative stability and are therefore superior cooking and frying oils. Palm oil available in various degrees of saturation serves as a good hard stock for the formulation of trans-free solid fat products. Blending palm oil with palm kernel oil or palm kernel olein and liquid oils such as super palm oil olein, sunflower oil or soybean oil produces solid fat products such as margarine/bread shortening, margarine/table spread, pastry/laminating fats with excellent functional, textural and storage properties (Dian et al. 2017).

As palm oil provides stability and economy to farmers, the world food chain has become highly dependent on palm oil. Since ancient times, palm oil has been incorporated into food, especially in Southeast Asia, where it was first discovered. However, as the food industry has become more modernized, palm oil has also increased in popularity on a global scale. The food industry benefits greatly from the characteristics of palm oil, including: high thermal stability, oxidative resistance, and low palm oil prices. Palm oil is therefore a desirable choice for use in various processed food products including deep-fried foods (Sumarlin et al. 2023).

Frying is one of the most popular food preparation techniques worldwide. Frying requires large amounts of hot oil in order to immerse the food with temperatures >100°C. During this process, many undesirable reactions occur that cause damage to the oil used. The reactions that occur in frying oil are mainly hydrolysis, thermal oxidation and polymerization (Setyawan at al. 2013). Therefore, the selection of food oil is important, so that the quality of fried food still meets human health.

According to Nor Aini and Miskandar (2007), the properties of cooking oil that must be considered are the ability of fat/oil to withstand high frying temperatures and be stable against oxidation and polymerization. The oil should have a high resistance to oxidation and gumming, a low increase in free fatty acids and smoke, a low foaming rate, a low darkening rate, a low melting point (except for special purposes) and a nutritious fatty acid composition. The saturated-unsaturated fatty acid content in the fat/oil determines the stability of the frying medium against oxidation.

Fats and oils containing high levels of saturated fatty acids such as animal fats result in high temperature and oxidative stability. However, such types of oils and fats are less preferred as they will cause health problems due to high consumption of cholesterol and saturated fatty acids. Whereas fats and oils containing high amounts of unsaturated fatty acids such as sunflower oil, soybean oil, and canola oil are prone to oxidation and produce undesirable odors and toxic compounds due to the breakdown of unsaturated fatty acids into di- and monoacylglycerols, glycerol, and free fatty acids (Zhao et al. 2023).

The study by Zhao et al. (2023) showed that at 52◦C palm oil samples were most stable against freeze-thaw compared to samples with other melting points. In texture and water holding capacity tests, the sample was able to increase the oil content from 5-20 g/100g, increase the gel strength and water holding capacity after freeze-thaw. Samples containing palm oil with the high melting point were less affected by temperature changes and showed higher stability detected by low-field Nuclear Magnetic Resonance and differential calorimetry scanning. Laser confocal scanning and rheology detection further confirmed the improved structural strength and freeze-thaw stability of emulsion gels containing solid fat particles with high melting point (Dian et al. 2017).

Palm oil contains significant amounts of vitamin E (tocopherols and tocotrienols), carotenoids, and, to some extent, phenolic compounds, as well as other bioactive compounds of considerable interest. Tocotrienols and β-carotene, along with other bioactive compounds in palm oil, are known for their healthful properties. These compounds were underutilized in many functional applications. Most of the literature on palm oil focus on the species *Elaeis guineensis*, but in the Americas, interspecific OxG hybrids-crosses of *E. guineensis* and *Elaeis oleifera*-have emerged. These hybrids have higher oleic acid content and richer bioactive compound profiles, expanding more research and application opportunities. Tocotrienols and β-carotene exhibit potential cardiovascular, neuroprotection, antioxidant, anti-inflammatory properties, and vision health benefits. Oils with higher oleic acid content are also noted for their health-promoting properties.

There are several challenges regarding the stability of the compounds during extraction, storage, and food processing that must be addressed. The review encourages further research into lesser-known compounds, such as squalene, phytosterols, and phenolics, and calls for deeper exploration into their health properties. Emphasizing sustainable extraction and encapsulation technologies offers an avenue to create pioneering products that improve health, meeting the growing consumer demand in this sector (Olivares-Tenorio et al. 2024).

In addition, red palm oil currently can also be utilized as a source of functional foods and nutraceuticals that are highly beneficial for human health. Phytonutrient compounds contained in red palm oil are β-carotene (pro-vitamin A), tocopherols and tocotrienols (vitamin E), phytosterols, squalene and ubiquinone. The content of oleic acid also has an impact on cardiovascular health, inflammation, microbes in the gut, neuroprotection, visual health, cancer, and bio accessibility (Hasibuan, 2021).

The Role of Women and the Economic Cycle in the Palm Oil Industry

The 2009 World Summit on Food Security defines global food security as the situation when all people, at all times, have physical, social and economic access to sufficient, safe and nutritious food to meet their dietary needs and food preferences for an active and healthy life (World Summit on Food Security (WSFS), 2009).

Tabe-Ojong et al. (2023) examined the relationship between palm oil production, and food security and diets in Cameroon and Indonesia, two palm oil producing countries with historical backgrounds of oil palm, as well as different cultivation, processing and marketing systems. The study found that (1) there is little evidence of a statistical relationship between oil palm production and food security in Cameroon and Indonesia; (2) while oil palm production is positively associated with dietary diversity in Indonesia, oil palm production has a negative relationship with dietary diversity in Cameroon. Oil palm farmers in Cameroon consume a less diverse diet than non-oil palm farmers. In contrast, in Indonesia, oil palm farmers perform better than non-oil palm farmers and consume a more diverse diet. This may explain why smallholders in Southeast Asia continue to adopt this crop rapidly; and (3) different pathways may explain the relationship between oil palm production, food security and dietary diversity in Cameroon and Indonesia. The findings also suggest that income, employment and diversity of agricultural production can explain the observed relationships between oil palm adoption, food security and diets. Given this, oil palm production may not be a universally suitable strategy to ameliorate food insecurity and nutrition but may be useful in some areas of production. Context-specific policies are needed to make oil palm cultivation and food systems more nutritionally sensitive and environmentally friendly.

In the plantation world there are several rewards that employees get for their work performance, including: salary, incentives, allowances and bonuses. Salary is a form of direct financial compensation. The salary is fixed, and the amount is certain as long as the employment relationship lasts. Salary is a monetary reward given to employees for the energy, thoughts that have been contributed in order to advance or realize company goals. Salaries are fixed and given in a fixed amount as long as there is a working relationship. For a harvest employee, salary has a deep meaning, which is something that can affect the level of life of the harvest employee concerned and his family. For this reason, the important thing in providing salaries is that the salaries received by harvest employees can provide encouragement or motivation to work hard, and motivation to live longer. Incentive is the results - oriented compensation work and is one factor important for motivating workers to be more enthusiastic in achieving its objective organization. Incentive is rewards paid directly to employees on performance accomplished work or exceeded the targets set determined.

Allowance can be another important consideration for employees in submitting their application to work in a company. The attractive allowance can make an employee more comfortable to work with. Allowance is in fixed amounts that are given continuously to the employees without any influences, such as employee's presence or other things. Employees that have a higher position in the office will get more allowance. In addition, there is a special allowance given every month from the company to cover water and electricity bills. The employees get more money during holiday seasons, such as Christmas or Eid celebrations. Bonuses are rewards received employees on achievement surpassing performance results. The actual bonuses are easy or not complicated to be calculated, because the bonus is measured based on the results that exceeded targets on a daily, weekly, monthly or annual basis.

The development and sustainability of oil palm plantations can create jobs and overcome the unemployment problem in Indonesia, which in August 2016 reached 7.03 million people (Ngadi and Noveria, 2017). The employment opportunities created in palm oil plantations are positively correlated to the land area, which reached 10.21 million ha in 2014. The land area consists of 4.98 million ha of private company plantations, 0.69 million ha of state company plantations, and 4.54 million ha of smallholder plantations. The potential to create employment opportunities and improve the welfare of workers in palm oil plantations is still large, especially in downstream products, because most of Indonesia's palm oil products are still sold in the form of raw materials (CPO).

The province with the largest number of palm oil farmers is Riau Province (526,350 households), indicating that most of the palm oil plantations are smallholder plantations. The land area managed by farmers in Riau Province in 2013 was 1.35 million ha, while the land area managed by state and private companies was 1.01 million ha with a labor absorption of 272,737 people. This land area makes Riau the province with the largest land area in Indonesia. The large number of oil palm farmers in Riau Province is inseparable from the welfare level of palm oil farmers which is quite good, thus encouraging other farmers to cultivate palm oil (Mehraban et al. 2022).

Palm oil is native to Cameroon but is an exotic crop in Indonesia that was commercialized several decades ago. Palm oil farmers in Cameroon consume a less diverse diet than non-palm oil farmers. In contrast, in Indonesia, palm oil farmers perform better than their non-palm oil counterparts and consume a more diverse diet, perhaps explaining why smallholder Southeast Asians continue to adopt the crop rapidly (BPS Musi Banyuasin regency, 2016). They showed that no statistically significant relationship was obtained between palm oil production and food security. The findings also suggest that income, employment, and agricultural production diversity might explain the observed relationship between oil palm adoption, food security and diet. Given this, palm oil production may not be a universally suitable strategy for ameliorating food insecurity and nutrition is useful in some production settings. Context-specific and tailored policies are needed to make palm oil cultivation and food systems more nutritionally sensitive and environmentally sustainable.

As the times progress towards the realization of gender equality, the role of women is increasingly recognized as equal. Many leaders, both in the economic, social and political fields, are held by women and prove their success. Reality proves that there is equality of roles in modern society where women are able to be equal and even superior to men in some aspects. Employment opportunities for women are also available in palm oil plantations, still considering the nature of women who experience menstruation, childbirth. Leave is obtained every month on the day (1-2 days) at the beginning of menstruation without reducing salary or cutting the leave period, as well as for the period of childbirth. Oil palm plantations in Indonesia still consider gender differences in terms of work targets, but with wages that do not differ between male and female workers.

Table 1. Daily time allocation of female and male adults by farm household type

Activities	Female			Male		
	(1) Only rubber	(2) Only palm oil	(3) Rubber and palm oil	(4) Only rubber	(5) Only palm oil	(6) Rubber and palm oil
Working on farm (hours)	2.74	1.50***	2.47	4.38	4.28	4.88
	(2.86)	(2.61)	(2.62)	(2.98)	(3.05)	(2.89)
Working off farm (hours)	1.37	1.29	1.08	2.18	2.24	1.71
	(2.66)	(2.76)	(2.60)	(3.20)	(3.29)	(3.09)
Household chores and care work (hours)	4.16	4.85*	4.38	0.43	0.52	0.51
	(2.83)	(2.69)	(2.59)	(0.92)	(1.18)	(1.52)
Leisure (hours)	3.34	3.96*	3.70	3.23	3.89**	3.33
	(2.28)	(2.52)	(2.49)	(1.90)	(2.35)	(1.95)
Eating and grooming (hours)	3.48	3.86*	3.61	4.42	4.20	4.32
	(1.49)	(1.93)	(1.78)	(1.43)	(1.46)	(1.55)
Resting/ sleeping (hours)	8.78	8.51	8.64	8.62	8.40	8.62
	(1.21)	(1.18)	(1.24)	(1.34)	(1.33)	(1.48)
Σ observations	268	94	173	264	96	179

Notes: Mean values of individual 24-hour time allocation data are shown with standard deviations in parents. Data from 2018 survey round. Mean difference tests carried out for columns (2) and (3) in comparison to column (1), and for columns (5) and (6) in comparison to column (4). *p < 0.05, **p < 0.01* **p < 0.001. Source Mehraban et al. (2022).

Using sex-disaggregated survey data from farmer households in Sumatra, Mehraban et al. (2022) assessed how palm oil cultivation - compared to traditional crop cultivation - is associated with time allocation and decision-making power for women and men. Women in palm oil cultivating households spent less time on farming and more time on household chores and leisure than women in rubber or other agricultural households (Table 1). Meanwhile, a reduction in women's workload can have positive social impacts.

Achieving zero hunger in Cameroon and Ghana requires gender-inclusive land tenure and labor policies that recognize, protect and empower smallholder farmers and women, and create opportunities for local and national food self-sufficiency (Fonjong and Gyapong, 2021). It was also found by Etuaha et al. (2020) that activities in plantations contribute to women's empowerment. Women with formal education and opportunities to access credit and formal training in agriculture and livestock generate more income and are more empowered.

The Palm Oil Industry: Developments and the Environment

Recent publications in various media outlets contain a long list of companies that have eliminated the use of palm oil. Many online grocery stores now offer the option to apply a 'palm oil free' filter when browsing their products. Why are consumers no longer using palm oil? And is this reputation justified? Some key questions about palm oil production: how has it changed, where is it grown, and how does it impact deforestation and biodiversity? The story of palm oil is more complex than it is often portrayed. In a large-scale UK consumer survey on perceptions of vegetable oils, palm oil was deemed the least environmentally friendly oil. 41% of people considered palm oil 'not environmentally friendly,' compared to 15% who considered soybean oil, 9% who considered rapeseed oil, 5% who considered sunflower oil, and 2% who considered olive oil. 43% also answered 'Don't know,' meaning almost no one considered it environmentally friendly (Fahamsyah, 2020).

Palm oil plantations are often criticized for their damage to the environment, by replacing tropical rainforests (Hansen et al. 2009), driving land cover and land use change (LCLU) (Xin et al. 2021), inducing CO_2 emissions (Guillaume et al. 2015), threatening biodiversity (Barnes et al. 2014), and harming other ecosystem services (Ganser et al. 2017). Palm oil has a high yield compared to other oil crops (FAO, 2015). This means that palm oil is likely to continue to play an important role in the near future, as the expansion of other oil crops could lead to more severe direct and indirect environmental and socio-economic impacts.

In Sumatra, where more than 60% of Indonesia's oil palm plantations are cultivated, smallholders currently control around 50% of the oil palm plantation area. But the rapid expansion of oil palm plantations is not without controversy. The palm oil industry recognizes several certifications, Indonesian sustainable palm oil (ISPO) and roundtable sustainable palm oil (RSPO). Indonesian sustainable palm oil (ISPO) certification, a system that strengthens and increases the acceptability of Indonesian palm oil products globally. The implementation of ISPO is stipulated in presidential regulation number 44 of 2022 concerning the certification system for Indonesian sustainable palm oil plantations. The certification system for sustainable palm oil plantations (ISPO) is mandatory or required for all palm oil plantation business actors (BUMN, private sector, smallholders) after the issuance of regulations governing it. There are 7 principles of ISPO certification listed in MOA 11/2015, including i) compliance with laws and regulations; ii) implementation of good plantation practices; iii) management of the environment, natural resources and biodiversity; iv) labor responsibilities; v) social responsibility and community economic empowerment; vi) implementation of transparency and vii) sustainable business improvement. This standardization accommodates government regulations ranging from land legality, waste handling to the welfare of company employees (Fahamsyah, 2020).

Ghana and Indonesia (Jelsma et al. 2019) conducted research to draw attention to an aspect of certification schemes that has not received much attention: their relevance to global and local food security. The RSPO and other voluntary certification schemes provide an interesting illustration of innovative instruments introduced to steer global food supply chains towards more attention to non-economic issues. Their impact on global and local food security has so far not been considered, and while the impacts are largely unintended, they may be considerable. Therefore, according to Jelsma et al. (2019) it is important to analyze and understand how voluntary sustainability governance mechanisms are shaped and what spillover impacts they have on food security. Government involvement is also important to complement private initiatives, or by encouraging more diverse and focused local interventions.

In Indonesia, this could be realized by policies that limit the amount of palm oil used to guarantee biofuels, e.g., cooking oil and biofuels for the domestic market. In Ghana, policies based on the conservation of agricultural biodiversity could encourage buying companies to diversify their product portfolios and reward a diversity of agricultural practices and entrepreneurial strategies based on dynamic social and agro-ecological environments.

In general, the case of the sustainability index assessment of palm oil-based bioenergy in Indonesia explains that its sustainability status is still low or less sustainable (Papilo et al. 2018). Of the three aspects (socio-economic and environmental), the sustainability status of the social aspect is still very low compared to the others. The main priority of the Indonesian Government is still more likely to lead to economic aspects. Therefore, bioenergy development requires more balanced attention to all the three aspects and each sustainability indicator. The economic results obtained from the development of palm oil-based bioenergy should be allocated and used to improve environmental and social aspects. This concern needs to be demonstrated by every stakeholder at every strategic level, from policy formulation to operational management in the field. RSPO stakeholders' perceptions of improving the overall sustainability of palm oil differ widely. Yet, there appears to be an underlying optimism among some RSPO stakeholders and local communities regarding the feasibility of technical improvements. Understanding RSPO stakeholders' perceptions can enrich the current debate on sustainable palm oil production and consumption.

Conclusion

To ensure the positive commitment of oil palm plantations to the economy, sustained efforts are needed to realize conditions that are environmentally friendly, quality and economical. The food industry is gradually shifting to economical palm oil sources and committing to environmentally and socially responsible food products. Activities such as power certification and government regulations can ensure that palm oil used in food products is made in a way that does not harm the environment. These certifications can also support the best skills in achieving a balance between financial benefits and maintenance of the natural environment.

Disclaimer

None.

References

Agricultural production statistics. 2000–2021.

Annual Report BPDPKS. 2021. Development source Power man plantation coconut palm as effort going to sustainable Indonesian palm oil. *Report annual Annual report*.

Ayompe, L. M., Nkongho, R. N, Masso, C Egoh, B. N. 2021. Does investment in palm oil trade alleviate smallholders from poverty in Africa? Investigating profitability from a biodiversity hotspot, Cameroon. *PLoS One* 16, 00256498.

Barnes, A. D., Jochum, M., Mumme, S., Haneda, N. F., Farajallah, A., Widarto, T. H., Brose, U. 2014. Consequences of tropical land use for multitrophic biodiversity and ecosystem functioning. *Nat. Commun.* 5(1), 1–7. https://doi.org/10.1038/ncomms 6351.

BPS Musi Banyuasin Regency. 2016. *Statistics area Subdistrict Tungkal Jaya in 2016*. Musi Banyuasin: BPS Kab. Musi Banyuasin.

Byerlee, D., Falcon, W. P., Naylor, R. 2017. The Tropical Oil Crop Revolution. *Food, Feed, Fuel, and Forests*. Oxford University Press, New York.

Central Statistics Agency. 2020. *Statistics coconut Indonesian palm oil.* https://webapi.bps.go.id/download.php?f=SRhdNH+fQ0szj1FJknsCM9GY8MhjFZi9qigGPb+GZ5lhJfBDByo0ohrmTAOGvh4XdsYBeDM1ZyG0T84bMWQUrt3E8ltGqVvmll7JZp4FrUvWAWwnlu58DftKl6aP/97z7RYphFpCPc4njEmZyQApcKbXOLHwW2T01k6kYKU2GxrlrAPRgGH62fJIK9J3X4gshLKO493JACSQ== [Accessed 18th April].

Dian, N. L. H. M., Rafidah, A. H., Kanagaratnam, S., Isa, W. R. A., Norazura, A. M. H., Ismail, N. H., Omar, Z., Sahri, M. M. 2017. Palm Oil and Palm Kernel: Versatile Ingredients for Food Applications. *Journal of Oil Palm Research* 29(4): 487–511 doi: https://doi.org/10.21894/jopr.2017.00014.

Directorate General of Plantation, 2014. Book publication 2014-2016. https://ditjenbun.pertanian.go.id/?publication=book-publication-statistics-2014-2016.

Directorate General of Plantation, 2022. Regulations minister agriculture republic Indonesia number 03 of 2022 concerning development source Power humans, research and development, rejuvenation, as well facilities and infrastructure plantation coconut palm https://ditjenbunppid.pertanian.go.id/doc/16/Permentan%20Nomor%203%20 Tahun%202022.pdf [Accessed 18th April].

Etuah, S., Ohene-Yankyera, K., Aidoo, R., Haleegoah, J., Wiggin, S., Henley, G. 2020. Impact of palm oil-related activities on women's empowerment in Ghana. *World Development Perspectives* 19: 100225.

Fahamsyah, E. 2020. *Opinion: Presidential Decree Number 44 of 2020 concerning System ISPO certification.* https://www.bpdp.or.id/opini-perpres-nomor-44-tahun-2020-tangan-sistem-sistem-ispo

FAO, 2015. FAOSTAT Databases. *Food and Agriculture Organisation*, Rome, Available at: http://faostat.fao.org/ (accessed 26.08.15).

Fonjong, L., Gyapong, A. Y. 2008. Gender roles and practices in natural resource management in the North West Province of Cameroon. *Local Environ.* 13(5), 461–475.

Ganser, D., Denmead, L. H., Clough, Y., Buchori, D., Tscharntke, T., 2017. Local and landscape drivers of arthropod diversity and decomposition processes in oil palm leaf axils. *Agric. For. Entomol.* 19(1): 60–69. https://doi.org/10.1111/afe.12181.

Guillaume, T., Damris, M., Kuzyakov, Y., 2015. Losses of soil carbon by converting tropical forests to plantations: erosion and decomposition estimated by δ13C. *Global Change Biol.* 21(9): 3548–3560. doi: https://doi.org/10.1111/gcb.12907.

Hansen, M. C., Stehman, S. V., Potapov, P. V., Arunarwati, B., Stolle, F., Pittman, K., 2009. Quantifying changes in the rates of forest clearing in Indonesia from 1990 to 2005 using remotely sensed data sets. *Environ. Res. Lett.* 4(3): 034001. doi: https://doi.org/10.1088/1748-9326/4/3/034001.

Hasibuan, H. A. 2021. Potency oil palm red as food functional and nutraceutica. *WARTA Research Center Coconut Palm Oil, 26* (3), 178-184. https://doi.org/10.22302/iopri.war.warta.v26i3.52.

Indonesian Oil palm research institute, 2016. History and development coconut palm oil in. Indonesia. National seminar history plantations in Indonesia. In frame inauguration of the plantation museum in Medan.

Jelsma, I., Woittiez, L. S., Ollivier, J., Dharmawa, A. H. 2019. Do wealthy farmers implement better agricultural practices? An assessment of implementation of Good Agricultural Practices among different types of independent oil palm smallholders in Riau, Indonesia. *Agricultural Systems* 170: 63–76.

Mehraban, N., Debela, B. L; Kalsum, U., Qaim, M. 2022. What about her? Oil palm cultivation and intra-household gender roles. *Food Policy* 110: 102276.

Ngadi., Noveria, M. 2017. Sustainability plantation coconut palm oil in Indonesia and prospects development in the region border. *Indonesian Society*, Vol. 43 No. 1, June 2017

Nor Aini, I., Miskandar, M. S. 2007. Utilization of palm oil and palm oil products in shortenings and margarine. *European J. Lipid Science and Technology, 109*: 422-432.

Olivares-Tenorio, M. L., Cortes-Prieto, C. M., Londono-Univio, N. C., Rojas-Diaz, D. A., Quintanilla-Carvajal, M. X., Tibaquira -Perez, L. N., Ricaurte -Puentes, L. Y., Rada-Bula, A. I., Romero, H. M., Garcia-Nunez, J. A. 2024. Bioactive compounds in palm oil: A comprehensive review of recent advances in physicochemical characteristics, health-promoting properties and technologies for extraction, concentration, fractionation, encapsulation and functional food applications. *Journal of Food Composition and Analysis* 132 :106306.

Original Source: The Star, 22 Dec 2017, (Website: http://thestar.com.my) ã Star Publications (M) Bhd 2017.

Papilo, P., Hambali, M. E., Sitanggang, I. S. 2018. Sustainability index assessment of palm oil-based bioenergy in Indonesia. *Journal of Cleaner Production* 196 :808e820.

Qaim, M., Sibhatu, K. T., Siregar, H., Grass, I., 2020. Environmental, economic, and social consequences of the oil palm boom. *Annual Review of Resource Economics* 12, 321–344.

Quality International. 2022. *CPO (Crude Palm Oil): Definition, Examples and Types.* https://mutucertification.com/cpo-pengertian-cepat-besar/.

Ritchie, H. 2021. *Palm Oil. Explore palm oil production across the world and its impacts on the environment.* https://ourworldindata.org/palm-oil.

Santika, T., Wilson, K. A, Budiharta, S., Law, E. A, Poh, T. M., Ancrenaz, M. 2019. Does oil palm agriculture help alleviate poverty? A multidimensional counterfactual assessment of palm oil development in Indonesia. *World Development*, 120, 105–117.

Setyawan, A. D., Sugiyarto, S., Susilowati, A. 2013. Physical, physical chemistry, chemical and sensorial characteristics of the several fruits and vegetables chips produced by low-temperature of vacuum frying machine. *Nusantara Bioscience* 5: 86- 103.

Sindonews.com. 2023. *Request Oil Vegetable Will Soaring so 307 million Tons, Opportunity For Indonesian CPO.* https://www.astra-agro.co.id/2023/06/14/permintaan-minyak-nabati-bakal-melonjak-jadi-307-juta-ton-peluang-bagi-cpo-indonesia/ Indonesia lifts temporary tires on palm oil exports. https://www.fao.org/giews/food-prices/food-policies/detail/en/c/1529567/ [Accessed 18th April].

Sumarlin, Sangkala, Zainuddin, Gaffar, U. H., Sutiharni, Husen, O. O., Robbani, S., Alam, R. A. C., Harwanto, F., Jayadisastra, Y., Pramulya, R., Anam, K. 2023. *Oil coconut and oil Palm Oil: Health, Environment, Economics and Social Behind Production.* CV Tohar Media, Gowa. ISBN: 978-623-8421-11-4.

Tabe-Ojong, M. P., Alamsyah, Z., Sibhat, K. T. 2023. Oil palm expansion, food security and diets: Comparative evidence from Cameroon and Indonesia. *Journal of Cleaner Production* 418: 138085.

World Summit on Food Security (WSFS), 2009. *Declaration of the World Summit on Food Security*, 16-18. FAO, Rome p. 2009.

Xin, Y., Sun, L., Hansen, M. C. 2022. *Land cover and land use change trajectory hopping facilitates estate crop expansion into protected forests in Indonesia.* Working Paper, available at SSRN: https://ssrn.com/abstract=3955794.

Zhao, D., Liu, S., Li, H., Yoong, J. H., Miao, Z., Cao, J., Liu, X. 2023. Effect of palm oil content and melting point on the freeze-thaw stability of fat substitutes. *LWT - Food Science and Technology* 188: 115452.

Chapter 11

Challenges and Prospects of Food Security in North Sumatra

Darma Bakti*
and Rosmayati
Department of Agrotechnology, Faculty of Agriculture, Universitas Sumatera Utara, Medan, Indonesia

Abstract

Food availability must be ensured year-round for social and economic stability in North Sumatra. This necessity was highlighted during two significant events: the communist rebellion in 1965 and the people's reformation of the government in 1998 when people were forced to take various kinds of food from market shops. Based on these experiences, the government should be responsible for maintaining adequate stock and stabilising prices to ensure they are affordable for the population. They should also implement effective policies and strategies to guarantee the availability and accessibility of nine essential food items: rice, sugar, eggs, meats, cooking oil, salt, chillies and onions. In 2022, the paddy field area in North Sumatra covered about 411,000 hectares, producing approximately 2 million tons of dried grain, which translates to 1.8 million tons of rice. With an average national consumption of 81.72 kg per person in 2022 and a population of around 17.2 million, North Sumatra required about 1.41 million tons of rice, leaving a surplus of 0.42 million tons. This data indicates that food security is currently well-maintained. However, the government must address the potential decrease in paddy field areas, which can be mitigated through subsidies for fertilisers and quality seeds, ensuring the Farmer

* Corresponding Author's Email: darma@usu.ac.id.

In: Natural Resources, Environment, and Food and Energy Security …
Editors: Mohammad Basyuni and Severino Salmo III
ISBN: 979-8-89530-283-5

Exchange Rate (FER) remains greater than 1. Looking ahead, it is crucial to encourage young and educated individuals to pursue careers in farming. Introducing appropriate technologies and innovations can help minimise input costs. Additionally, government bodies, particularly the Board of Agriculture and financial institutions such as banks, should reward and appreciate the young generation's involvement in agriculture.

Keywords: food security, farmers, rice production, government, technology

Introduction

Food plays a crucial role in human life. Rice is a strategic staple food in Indonesia. In fact, the main food in all ASEAN countries also comes from rice. In past decades, some regions had staple foods other than rice, such as Madura, which consumed corn and Papua, whose main food was sago. With the government's policy to demonstrate equitable development, a campaign was launched to promote the health benefits of rice-based foods. To show the seriousness of equality among all citizens in development, the government encouraged all regions in Indonesia to eat rice. Additionally, there was an assumption that staple foods other than rice implied a lower or backward social status. To eliminate this perception, the government aimed to demonstrate that equitable development had taken place by ensuring that the entire country adopted rice as their staple food. Indonesia is geographically divided into three regions: west, central and east. Since the implementation of this policy, Indonesians from Sabang to Merauke have adopted rice as their primary food source (Nasution, 2019). However, this shift has increasingly complicated the government's efforts to meet the nation's food needs over time. The continuously growing population and unhealthy consumption patterns have made it challenging for the government to ensure adequate food supply for the population (Nasution, 2021). Despite persistent efforts to achieve food self-sufficiency, the government faces significant challenges, including unpredictable climatic conditions, reduction of arable land, population growth, low technology utilisation, limited human resources, and economic factors such as farmers' exchange rates. To address the food shortage experienced from early January to February 2024, the government was compelled to import 3.6 million tons of rice from rice-producing countries, with plans to import up to 4.1 million tons if the situation deteriorates further (Pramono et al. 2020). In Indonesia's development agenda, particularly in agriculture, the concept of sustainable development is pivotal. Sustainable development aims to meet current needs without compromising the ability of future generations to meet their own needs. This approach encompasses three key aspects: economic, environmental, and social sustainability.

National Library's (2022) national development or the Millennial Development Goals (MDGs) encompass 17 goals, with at least two directly related to daily life: 1. freedom from hunger and 2. healthy living (Djaiman et al. 2018). The quantity and quality of food are crucial for achieving a healthy life. Healthy food produces an intelligent generation capable of competing with other nations. The tendency for rice consumption in Indonesia is quite high compared to neighbouring countries. Therefore, an appropriate solution is needed to address food shortages through stages of agricultural development, starting from food security, then moving towards food self-sufficiency, and finally achieving food sovereignty by 2045.

Prospects of Food in Indonesia

The national rice production in 2020 amounted to 55.16 million tons of dry milled unhulled rice (GKG), an increase of 556.51 thousand tons or 1.02% compared to the 2019 production of 54.60 million tons of GKG. If the potential rice production in 2020 is converted to rice, it becomes 31.63 million tons. This data shows an increase of 314.10 thousand tons or 1.00% compared to 2019, which was 31.31 million tons (BPS, 2020). Meanwhile, the rice production in 2023 was 53.63 million tons of GKG, a decrease of 1.12 million tons or 2.05% compared to the previous year. Rice production in North Sumatra was 2.09 million tons, equivalent to 1,186,348.84 tons of rice, with an area of 411.46 thousand hectares. Given that the population of North Sumatra Province is 14,936,148, the rice demand is 1,957,882 tons. This indicates a rice deficit in North Sumatra. The harvested area of rice in North Sumatra Province in 2022 was 413.140 thousand hectares, an increase of 4,964 hectares or 1.22% compared to 2021 (BPS, 2021).

If the rice production in 2022 is converted to rice for consumption, the rice production in North Sumatra Province in 2022 would be 1,186,350 tons, a decrease of 16.77 thousand tons or 1.33% compared to 2022. The population of North Sumatra is currently 17,372,437 based on the 2022 population census, an increase of 1.82 million people (BPS, 2022).

According to the above data, the population growth rate of North Sumatra is 1.28% per year. The prospect of food security in North Sumatra remains good, using various strategies such as opening new rice fields and stopping the continuous conversion of land, especially land close to cities. Additionally, the government needs to implement specific policies to promote food diversity. Many agricultural products such as tubers, corn, bananas and breadfruit, which grow easily and are abundant in North Sumatra, can replace rice as the main food ingredient and are of comparable quality to rice.

Indonesia boasts the richest biodiversity in the world, with the capacity to cultivate almost all types of plants found in Latin America, Africa and even Europe. Indonesia's climate, characterised by only two seasons, the dry season and the rainy season, contributes to the land's exceptional fertility. In certain regions, subtropical plants such as cabbage, mustard greens, onions, tomatoes and potatoes can be successfully grown and thrive. Consequently, concerns about food shortages can be mitigated, provided that planning and implementation are conducted properly and sustainably (Restra, 2020).

Challenges

The issue of food, particularly rice, has been a national problem from one government to the next. This was highlighted in President Sukarno's speech on 27 April 1952, during the laying of the foundation stone for the Faculty of Agriculture at the University of Indonesia, located in Bogor. The Faculty of Agriculture later evolved into the Bogor Agricultural Institute on 1 September 1963. At that time, President Sukarno emphasised that Indonesia faced numerous challenges that needed to be addressed to transform the nation into an advanced country. One of the most pressing issues was the threat of hunger. According to statistical data from 1940, Indonesia's population was 75 million, with an average rice consumption of 86 kg per capita. The nation's food requirements included not only rice but also tubers, corn, soybeans and other staples. It was estimated that 6.5 million tons of rice were needed, while production was only 5.5 million tons, resulting in a shortage of 1 million tons that had to be imported (Agriculture Development polytechnic, 2019). Sukarno noted that in 1940, the daily food requirement per person was 1,700 calories, but by 1952, it had increased to 1,850 calories per person. The issue of achieving food self-sufficiency remains unresolved even today. In Medan, rice consumption levels are particularly high: for low-income households, the average is 40.2 kg per month; for middle-income households, 50.4 kg per month; and for high-income households, 60.2 kg per month, with an average household size of four people. This data suggests a tendency for rice consumption to increase by an average of 10 kg per household as income decreases (Silalahi et al. 2020).

In a country known for its fertility and suitability for agriculture, both because of its tropical location and its fertile soil, one can still find communities experiencing insufficient food intake (malnutrition). The lack of adequate food intake leads to various health problems. Humans need food to obtain calories and energy to live normally and healthily. Without energy, the cells in the body cannot function properly, and eventually, the organs, including the heart and brain, cannot perform as they should. With too low calories intake, the organs lack the energy to perform their functions. As a result, there are disruptions in the body's functional balance, which can lead to various diseases. On average, adult men need around 2,500 calories a day, while adult women usually need around 2,000 calories. Insufficient calorie intake causes the organs to lack the energy to perform their functions, leading to disruptions in the body's functional balance and potentially resulting in various diseases.

According to FAO (1996) and Law No. 7/1997, four components must be met to achieve food security:

1. Sufficient food availability
2. Stability of food supply without seasonal or annual fluctuations
3. Accessibility to food
4. Food quality and safety

The lack of food intake can have mild, moderate or even severe effects depending on the extent of the deficiency. The problem is that the fulfilment of basic needs, namely rice, is still not resolved, let alone the fulfilment of nutritious food needs, especially during infant growth. The national average index for underweight, stunting (short stature) and wasting (thinness) due to malnutrition is 2.5, while in North Sumatra, it is 2.7 (Hastuti et al. 2018). It suggests that the issue of malnutrition needs to be a major concern for the North Sumatran government and higher education institutions. Three key factors significantly impact the Human Development Index (HDI): 1. Health, 2. Education and 3. Expenditure. The type of food consumed is closely related to health. If the food consumed is proportional and contains the necessary nutrients for the body, then most health-related issues will be addressed.

Food Development Plan

For a long time, the government has devised schemes to address food issues, particularly rice, by improving conventional methods to mechanised methods known as Modern Agriculture (Zulfahrizal, 2012). The modern approach involves expanding and renewing land and implementing mechanisation from the field process to post-harvest. However, these efforts have not yet shown satisfactory results since the characteristics of Indonesian food agriculture differ from those of neighbouring countries like Thailand. Many challenges are faced in maintaining food security to increase production. These challenges include the availability of seeds and production facilities, such as fertilisers and medicines, shrinking rice fields due to ineffective land protection laws, unstable agricultural human resources, as well as fluctuating selling prices of produce. These issues have not been resolved due to the involvement of numerous bureaucracies responsible for addressing them. In addition to direct assistance from the central government to farmers, there is also a lack of coordination among local governments at the provincial and district/city levels (Department of Food Security, 2016). Coordination between different levels of government, each with its authority, and program synchronisation as outlined in their respective RPJMs (Regional Medium-Term Development Plans) are crucial. This lack of coordination often hinders the effectiveness of aid in achieving its goals.

Food self-sufficiency in Indonesia has not been fully maintained, which is evident from the trade balance from 2017-2020, which shows both imports and exports of rice (Sabrina et al. 2022). However, the value of exports is much smaller than the amount of imports, indicating that exports should not occur as the demand for rice continues to rise. Since Indonesia's independence and throughout the reform era, food security has been a primary concern for the government. Nevertheless, the government's ongoing difficulties in providing an adequate supply of rice remain far from expectations. The government has created a roadmap for agricultural food development, particularly up to 2045 (Ministry of Agriculture, 2016).

During the Jokowi-Ma'ruf administration period of 2019-2024, the Ministry of Agriculture's RPJMN has been set to make Indonesian agriculture advanced and self-sufficient. Nationally, the goal for food agriculture over five years is to increase production and productivity. To achieve this productivity, mechanised tools and supporting research are needed, along with low-cost agriculture and the expansion of cultivated areas. The development of food agriculture in North Sumatra has not yet been optimal. Two key factors have not functioned well: 1. Agricultural resources or farmers, and 2. Production facilities and support infrastructure, such as farmer institutions and extension worker human resources. Primary and supporting facilities require accurate data and good coordination to achieve the planned targets. Indonesia's food security index is 62.2, below the average for countries in the Pacific region, which is 63.4. This places Indonesia at 63rd out of 113 countries. In comparison, Singapore has an index of 73.1, followed by Malaysia with 69.9. Singapore does not have agricultural land, yet it is secure from food shortages. Both countries have smaller populations than Indonesia. Additionally, as an archipelagic country, Indonesia faces longer distribution times and higher costs for transporting necessities, including food. From the above explanation, it can be concluded that the prospect of achieving food security is attainable. With sincere planning and consistent implementation, Indonesia can achieve food self-sufficiency and sovereignty by 2045 (Sibuea, 2021).

Conclusion

1. North Sumatra and South Sumatra are the largest rice granaries outside of Java. In North Sumatra, the biggest challenge in increasing production is the conversion of land to oil palm plantations and residential development. In addition, existing irrigation systems need continuous quality improvement to ensure adequate water supply for the rice fields.
2. To prevent farmers from switching to more promising jobs, the government must ensure that the difference between the selling price and the production cost is significant (farmers' exchange rate or NPT > 1). During harvest, the government can intervene through price-supporting institutions, including state-owned enterprises like Bulog, to purchase the harvest at fair prices, thereby improving farmers' welfare.
3. The government should increase aid or subsidies for essential materials, such as fertilisers, pest and disease control chemicals, high-quality seeds, and land processing tools, to farming groups or

associations. This will ensure that the government's support is genuinely felt by farmers when they face difficulties in managing their land.

Disclaimer

None.

References

Agricultural Development Polytechnic. 2019. [*A Matter of Life or Death.* President Sukarno's speech at the groundbreaking of the Faculty of Agriculture building at the University of Indonesia in Bogor on 27 April 1952].

BPS, 2020. *Sumatera Utara dalam Angka.* Badan Pusat Statisitk Sumatera Utara.

BPS, 2021. *Sumatera Utara Dalam Angka.* Badan Pusat Statistik Sumatera Utara.

BPS, 2022. *Sumatera Utara Dalam Angka.* Badan Pusat Statitik Sumatera Utara.

Centre for Agricultural Extension, 2014. [*Guidelines for the implementation of the Classification of District Extension Centres*]. Centre for Agricultural Extension. Badan Penyuluhan dan Pengembangan SDM Pertanian.

Darsono, 2012. [The Food Revolution starts with a Revolution in the Way of Thinking about Food. In *Food for the People: A Matter of Life or Death 60 years later*]. Publisher Department of Agribusiness, FEM IPB, Bogor Reflections on Bung Karno's Speech at the Groundbreaking of IPB Baranangsiang Campus, Bogor.

Department of Food Security Food Crops and Horticulture, 2016. [*Development of Food Agriculture and Horticulture in North Sumatra. Presentation of the head of the Food Crops and Horticulture Office of North Sumatra Province*]. delivered in Medan on 14 November 2017 to the Academic Community of the USU Faculty of Agriculture, Medan.

Djaiman, S. P. H., Nunik K. W., Sihadi, K. Sari. 2018. [Disparities in Undernourished Children in Indonesia]. *Health Research and Development Media.* 28(3). Kementerian Kesehatan. Republik Indonesia.

FAO, 1996. *The state of Food: Food: Macroeconomic dimension: Food and Agriculture Organization of the United Nation.*

Hastuti, S. P., Wardani, N. K., Sihadi, Sari, K. 2018. [Disparities in Undernourished Children in Indonesia]. *Health Development Research Media*, 28 (3). 201-210.

Law No 41 Tahun 2009. *On the Protection of Food Agricultural Land in Indonesia.*

Ministry of Agriculture, 2016. [*The Road Map Direction of Rice Commodity Development in Indonesia. delivered by the Minister of Agriculture at the Seminar of the Association of Agribusiness Students and Alumni*].

Ministry of Research and Technology of the Republic of Indonesia. 2006. [White Paper. Indonesia 2005-2025]. *Research and Application of Science and Technology in Food Security.*

Nasution, D. B. 2019. [*Food Agriculture Development in the Era of Industrial Revolution 4.0*]. Thoughts of USU Professors of Higher Education in the Era of the Industrial Revolution 4.0. USU Press Publisher, 2019.

Nasution, D. B. 2021. [Food Security Stability in Supporting Food Security in Supporting Public Health in North Sumatra]. *Professor's Thoughts*. Universitas Sumatera Utara. *USU Press.*

National Library, 2022. [*North Sumatra builds food security towards a healthy, active and productive society*]. USU Press.

Pramono, S., Supriana, T., Kesuma, S. I. 2020. [Analysis of Land Use Change and Prediction of Rice Production in Deli Serdang Regency]. *Media online Repository USU.*

Regulation of the Minister of Agriculture of the Republic of Indonesia No.28/Permentan/OT.140/4/2012 on guidelines for the Assessment of Outstanding Kecematan Agricultural Extension Centres.

Restra Ministry of Agriculture, 2020. [S*rategic Plan for Indonesia's Food Agriculture Development 2020-2024. Directorate General of Food Crops*].

Sabrina, R., Sirait, B. I., Manurung, A. I. 2021. [*Building Indonesia's Food Security from North Sumatra*]. USU Press.

Sibuea, P., 2021. [*Food Security. Realising Food Independence and Sovereignty*]. An Idea and Thought. Publisher Unika Press.

Silalahi, N. E., Salmiah, Jufri, M. 2020. [Consumption level and pattern of rice consumtion in Medan City]. *Repository USU.* Medan.

Zulfahrizal, 2012. [*Indonesian Rice Farming (Problems and Solutions) in People's Food: A Matter of Life or Death 60 years later*]. Publisher Department of Agribusiness, FEM IPB, Bogor Reflections on Bung Karno's Speech at the Groundbreaking of IPB Baranangsiang Campus, Bogor.

Chapter 12

Food Estate Policy Implementation in North Sumatra

Noverita Sprinse Vinolina[1,*]
and Tavi Supriana[2]
[1]Department of Agrotechnology, Faculty of Agriculture, Universitas Sumatera Utara, Medan, Indonesia
[2]Department of Agribusiness, Faculty of Agriculture, Universitas Sumatera Utara, Medan, Indonesia

Abstract

Indonesia is an agricultural country, with most of its population working in the agricultural sector. The population increase that occurred in Indonesia in 2020 exacerbated food insecurity and disrupted the food supply chain. Developing the food and agricultural systems can ensure sufficient, affordable, nutritious, and diverse food for all societal levels. The Food Estate (FE) policy, initiated by the Government of Indonesia, aims to address food security and food crises both in Indonesia and internationally. Despite the implementation of several large-scale food estate policies and programs, the FE policy has not yielded optimal results. The main hindering the success of FE practices in Indonesia include unsuitable development locations, limited water supply, inadequate technology, insufficient farmer cooperation, land tenure conflicts between the community and government, the limited capacity of farmers to manage agricultural land effectively and maximize production yields and quality. However, the FE policy has the potential to become a flagship initiative for the Government of Indonesia by focusing on environment sustainability, preventing deforestation,

* Corresponding Author's Email: noverita@usu.ac.id.

In: Natural Resources, Environment, and Food and Energy Security …
Editors: Mohammad Basyuni and Severino Salmo III
ISBN: 979-8-89530-283-5

maintaining biodiversity, and preserving water catchment areas. For instance, several strategies and modalities could be implemented to improve the FE program in North Sumatra, particularly regarding farm sustainability.

Keywords: agricultural sustainability, food estate, food security, North Sumatra

Introduction

The food problem in Indonesia is partly attributed to significant population expansion. High population growth increases food demand, leading to shortages. With roughly 270 million people, Indonesia is the fourth most populous country in the world and is considered as a lower-middle-income country. In addition to significant population growth, agricultural production in Indonesia is affected by insufficient agricultural land and limited access to adequate water sources due to climate change, ultimately leading to food insecurity and crises, resulting in malnutrition (Neilson and Wright, 2017; Dinanta et al. 2021).

The original objectives for developing the national Food Estate (FE) were to reduce reliance on imported food, predict changes in climate components, and prepare for a potential food crisis brought on by the COVID-19 pandemic. Food and agricultural development can supply all societal levels with an adequate, reasonably priced, safe, diversified, and nutritious supply of food. FE programs aim to implement a system of food security that is affluent, autonomous, sustainable, and independent for the people. These priority programs support food security through six key actions that enhance food quality, availability, and access. The agricultural sector is expected to generate high-quality food and agricultural products and gain international competitiveness with the help of the Sustainable Development Goals (SDGs) and advances in human and agricultural resources and technological innovation. This development aims to establish Indonesia as a significant processed food producer (Basundoro and Sulaeman, 2020; Juhandi et al. 2023). As part of its Vision 2045, the Indonesian government plans to develop a FE project to supply national food reserves and prepare for potential food crises related to COVID-19. According to Presidential Regulation Number 109 of 2020, regarding the Third Amendment to Presidential Regulation Number 3 of 2016 concerning the Acceleration of the Implementation of National Strategic Projects in the Era of President Joko Widodo, this FE project is part of the 2020–2024 National Strategic Program Plans (Rasman et al. 2021; Kompas, 2021).

The national FE was established in multiple locations, including North Sumatra, to enhance agricultural production areas and become a key component of the national food production chains. One such location is Humbang Hasundutan, North Sumatra, where commodities such as potatoes, shallots, and garlic are produced (Sigalingging et al. 2023a; Juhandi et al. 2023; Sigalingging et al. 2023b). However, the FE program in this region has not achieved optimal results due to several challenges: poor accessibility to the area, especially during the rainy season; limited farmer involvement in FE development process; unresolved community land ownership issues; and the use of deforestation practices in agricultural mechanisms.

The Potential and Constraints of Food Estate in Indonesia

The agricultural sector plays a significant role in the Indonesian economy, providing income for a large portion of the population through its abundant natural resources spanning from Sabang to Merauke. These resources include various plantation crops, such as oil palm, rubber, coffee, sugar cane, and tea. Oil palm, in particular, is one of the largest contributors to foreign exchange earnings. Rubber plantations flourish in Indonesia's climate. In addition, coffee, appreciated worldwide, offers diverse flavours across different regions, whether robusta, arabica, or liberica. Sugarcane and tea also feature prominently among Indonesia's plantation crops, further contributing to foreign exchange revenues (Dinanta et al. 2021; Saputra, 2021).

The concept of FEs, focusing on horticultural crops, has been developed since the old order period in 1955. The purpose of implementing FEs in Indonesia is to increase agricultural production and reduce reliance on imported products. Furthermore, FEs aim to foster independent food self-sufficiency through a food sovereignty approach rooted in ecological agriculture (Zannati, 2020). Despite its potential, the realization of FE projects in Indonesia faces numerous challenges, hindering significant success. Factors such as technological requirements, suitable locations, water availability, climatic conditions suitable for them, and synergy between the government and farmers, are crucial considerations in planning and implementation. Land ownership needs to be addressed for supporting the success of FEs in Indonesia as conflicts between the government, farmers, communities and entrepreneurs still persist (Yestati and Noor, 2021; BBC, 2021). Improving farmer skill is imperative to ensure effective management of agricultural land and mitigate factors leading to unsatisfactory agricultural outcomes. Government policies and regulations are crucial for the success of FEs in Indonesia, focusing on protecting local markets in order to increase agricultural productivity, the quality of human resources, and strengthening coordination among stakeholders regarding agricultural development (Sianipar and Tangkudung, 2020). Environmental issues such as carbon release due to deforestation, biodiversity loss, reduced water catchment areas, and flood disasters, also demand attention (Rasman et al. 2021).

Conducting research and feasibility studies, including in-depth environmental, economic, and social analyses, is crucial for the success of these food estate development projects. This ensures suitability to soil and climatic conditions, considers geographic features, social dynamics, regional diversity, and accessibility limitations (Puspitojati, 2015; Juhandi et al. 2023). Involve local communities in the project planning and implementation process to foster their ownership and commitment to proper land management. Deploy suitable technology and agricultural support facilities to enhance farming efficiency and boost food production. Promote agricultural knowledge through collaborative research and development initiatives involving universities and the private sector. Develop policies addressing global challenges such as rising food prices, population growth, and climate change, with the aim of increasing agricultural production efficiently and effectively through sustainable agricultural practices (Sianipar and Tangkudung, 2020).

Discussion

The FE policy, a national strategic program aimed at achieving food (Astuti, 2020), integrates various agricultural sectors and utilizes underutilized or degraded forest areas. The program comprehensively integrates food crops, horticulture, plantations, livestock, and fisheries. By combining forest plants with food crops through agroforestry and integrating forest crops with livestock, the program enhances the functionality of protected forests.

Despite forest areas defined as certain areas designated and/or determined by the government to maintain permanent forest status (Kementerian ESDM, 1999), they contain areas that are no longer forested.

These areas present opportunities for the development of FEs. With only 21% of the 30% non-forest area available for agriculture, Indonesia faces the challenge of meeting the food needs of its approximately 270 million people, a number projected to increase annually (Nurrochmat, 2020). Indonesia's limited agricultural land prompts the utilization of forest areas for agricultural activities.

The significant role of human-forest interactions, spanning from the beginning of civilization to modern times and projected into the future, is evident in five functional periods:

1. the era of complete human dependence on forests,
2. the phase of controlled collection of forest products,
3. the period of human-induced forest degradation,
4. the era where forests are essential for human survival, and
5. the age characterized by a growing appreciation and desire for forests.

The utilization of forests as a food source differs between countries, especially between developed and developing countries (Puspitojati, 2015). Developed countries convert most of their forests into agricultural land and other activities and allow the remaining forests to remain as forests. Meanwhile, in Indonesia, which has limited agricultural land, forests continue to be a source of food for people living in and around forests (Mutia et al. 2022). Cultivation of food crops in forests has its origins in shifting cultivation, which is closely related to cultural traditions in Asia. Forests are cleared for fields, planted with food crops for a few years and abandoned after the fertility of the land declines. Cultivation resumes in a new area and continues on a shifting basis until returning to the original area. Rotation cycles can be more than 30 years, depending on the ability of the land to recover. In line with population growth and the increasing need for land, available forest areas for farming diminish. Farmers are forced to cultivate in narrower areas or with shorter rotations. As a result, soil fertility declines, and the yield is reduced. To compensate, farmers expand their cultivated fields, further shortening their rotation cycles.

Continued forests destruction poses significant adverse effects, particularly for farmers. To mitigate these effects, farmers have experimented with various plant species to accelerate the recovery of land fertility. At first, farmers used trees to restore land fertility. However, trees could not completely restore soil fertility, leading the farmers to try additional external inputs such as fertilizers and pesticides to continue growing food crops. As a result, farmers no longer practiced shifting cultivation but settled farming, which transforms the forest areas into agricultural land.

The integration of FE development with forest areas for food security allows forest areas to be used for food estate projects. In temperate countries, large-scale conversion of forests to agriculture (food and non-food) is limited to farmlands. Forests are managed for timber, and food crops are separated from forest products. In developing countries like Indonesia, the strict division between agricultural and forestry crops does not apply to rural communities, especially those living in and around forests. There is cultivation of various types of timber, food and non-food crops within the forest. Various types of food crops are cultivated in the forest, such as rice, corn, cassava, sago, soybeans, porang and peanuts, fruits such as jackfruit, mango, avocado, mangosteen, melinjo, petai and jengkol, and industrial plants, such as coffee and vanilla. These food crops have been cultivated in plantation forest areas, namely through community forest (CF), industrial plantation forest (IPF), community plantation forest (CPF), plantation forest with non-timber forest products (NTFP), and village forest schemes. This shows that plantation forests have been designed to be managed for food, both periodically and sustainably, in order to improve the welfare of forest communities and the success of forest management (Puspitojati et al. 2014).

The basic concept of FE is the integration of sectors and subsectors in an agribusiness scheme with the optimal and sustainable utilization of resources managed professionally, and supported by quality human resources, environmentally sound appropriate technology, and strong institutions (Balitbangtan, 2011). FEs are directed towards agribusiness systems that are firmly rooted in rural areas based on the empowerment of indigenous or local communities which are the foundation of regional development. Notably, FEs can be carried out by individuals or business entities. In the event that it is carried out by a business entity, the government directs that each business cooperates in an integrated manner in carrying out the crop cultivation business.

The current redevelopment of the FE program by the Government of Indonesia is in anticipation of the food crisis predicted by the World Food Agency. The FE program is expected to become a pillar of national food security, contributing to economic stability, politics, and national security (Sianipar and Tangkudung, 2020). Currently, the FE program has been included as part of 201 National Strategic Projects regulated in Presidential

Regulation Number 109 of 2020 concerning the Third Amendment to Presidential Regulation Number 3 of 2016 concerning the Acceleration of the Implementation of National Strategic Projects.

Since the FE program is a large-scale food business initiative, it requires a tremendous amount of land, prompting the government to establish an appropriate system for allocating land in forest areas. In essence, the goal of the forest area project for food security is to ensure that there is land available for FE initiatives. Indonesia's commitment to lowering greenhouse gas emissions in the fight against climate change depends in part on the government's role in managing and preserving protected forest areas. The COVID-19 pandemic exacerbated the food insecurity in Indonesia. Specifically, given the various geographical, social, and economic circumstances in Indonesia, the food crisis situation varies between regions and community groups. COVID-19 interrupted the food supply chain and limited the availability and accessibility of food, caused unpredictable price swings, and triggered farmer work stoppage (Rozaki, 2020). The pressure on Indonesia's food security is also caused by declining community income, reduced access, and increasing transaction costs (Virananda and Ikhsan, 2021). To ensure that every person and household has access to enough food from the country's food production, food security must be increased by ensuring the availability of safe, equitable, reasonably priced food across Indonesia (Suharyanto, 2011).

The Indonesian government introduced the FE Policy to improve food security in the nation. A FE is a sizable agricultural area developed with the goal of supplying the local populace with food. However, FE development is expected to worsen Indonesia's food shortages and lead to a food crisis (Vinolina, 2020; Lasminingrat and Efriza, 2020). The 13th Sustainable Development Goal (SDG), which includes combating climate change, supporting sustainable agriculture, eradicating hunger, and ensuring food security and decent nutrition, is being worked toward by the Indonesian government (Stafford-Smith et al. 2017).

One of the laws that supports the realization of goals 2 and 13 of the SDGs is Law of the Republic of Indonesia No. 18/2012 on Food. Article 12, paragraph 5 of Law of the Republic of Indonesia No. 18/2012 on Food states is to realize food availability through domestic food production by: (a) developing food production based on local resources, institutions, and culture; (b) developing the efficiency of food business systems; (c) developing facilities, infrastructure, and technology for efficient food production, post-harvest handling, processing, and storage; (d) building, rehabilitating, and developing food production infrastructure; (e) maintaining and developing productive land; and (f) establishing food production centers.

Food insecurity can occur due to increased food demand that the food production system is unable to support. Food crises can caused as a consequence of natural (such as natural disasters, extreme weather, water crisis, land degradation) and non-natural factors (such as economic policies, political changes in a country, wars between countries, acts of terrorism, acts of separatism, global health and security conditions, biofuel alternative energy trends and dictatorial and kleptocratic governments) (Kamil and Elwan, 2021).

Based on the Presidential Regulation of the Republic of Indonesia Number 85 of 2021 concerning the Government Work Plan for 2022, FE development remains a strategic step to support economic recovery in Indonesia through increased food security. FE development, which has started in 2021, was implemented again in 2022 with priority locations in Central Kalimantan and North Sumatra Provinces. FE development can also diversify the economy by increasing added value, namely through increasing the contribution of the processing industry and the development of agricultural products.

In 2021, Humbang Hasundutan, Sumatera Utara, cultivated potato, onion, and garlic commodities under the National Food Estate Program. However, the project faced several limitations:

1. Steep and hazardous accessibility conditions; especially during the rainy season.
2. Lack of farmer involvement in the food estate development process.
3. Ongoing landownership issues among residents.
4. Challenges associated with investment-driven agricultural practices contributing to deforestation.

The government continues to evaluate and refine the Food Estate Program to meet its objectives by:

1. Implementation of a national data system to determine the production of each commodity in each region. There is a lack of agricultural infrastructure, long supply chains, expensive logistics costs, asynchronous availability of goods between regions which results in food distribution problems in Indonesia.

2. Developing commodities in the FE program to be more diverse and adapted to each region.
3. Implementing special technology to predict climate patterns and adjust planting strategies accordingly. The potential utilization of agricultural technology on FE land needs to be developed by utilizing greenhouse technology with an integrated system to control temperature, humidity, irrigation, nutrition, and post-harvest quality.
4. Fostering farmer corporations in FE programs. The community should be empowered as producers, distributors, and consumers.
5. Regulating the agricultural trading system and determining the crop types.
6. Setting price floors on agricultural products, so that the selling price cannot be below the floor price.
7. Integrating agricultural technology research, technology funding, and food technology research to maintain production stability and sustainability of the agricultural sector. This can be achieved by encouraging productivity through advanced agricultural technology, such as seeds, fertilizers, and tractors. Technological development needs to be synergized between the government, universities, and the private sector.
8. Creating policies for providing direct payments to any farmer willing to voluntarily join the demand-supply adjustment system. Thus, the government can ensure that farmers' welfare is maintained.
9. Transforming agrarian policy (targeting 50% food self-sufficiency and maintaining multifunctional land)
10. Changing agricultural policy orientations from reduction patterns to production systems.

Enhancing the North Sumatra Food Estate Program: Three Strategies

First Strategy: Financial Enhancement

The first strategy aims to enhance financial aspects through the organization of agricultural capital, improvement of industry accessibility, and promotion of agricultural insurance. The implementation techniques for the proposed measures are outlined below:

1. Cultivating capital: The government encourages agriculturists/ farmers to borrow capital from banks at reduced interest rates. Additionally, the government regulates credit repayment schedules in alignment with the harvest season.
2. Industry accessibility: Collaboration between the food industry and the government is fostered. Incentives such as corporate tax deductions and opportunities for participation in land processing at FE sites, are provided to the industry. Locales are classified by the government based on product costs, and crops can be showcased or disseminated to other locales through Bulog (the National Logistic Agency). The government also supports the launch of small-scale processing businesses through the introduction and adoption of agricultural processing technology, empowering local communities.
3. Promotion of agricultural Insurance: The government promotes the use of agricultural protections services by insurance companies and agriculturists. Tax deductions for companies are provided to facilitate the provision of agricultural insurance services to FE farmers. Moreover, the government encourages FE farmers to utilize insurance services by disseminating information on the benefits of insurance and subsidizing the initial payment of insurance premiums.

The provision of agricultural insurance mitigates potential losses resulting from crop failure, which can occur as a result of different variables. The establishment of processing industries provides farmers with a viable market to sell their produce, reducing reliance on intermediaries. Financial institutions serve as a source of capital for agriculturists to scale up their agrarian commerce (Saputra, 2021; Juhandi et al. 2023).

Second Strategy: Culture Advancement

Culture advancement involves improving social perspectives, such as communal work, agriculturists orientation, and agricultural cultural information. This pproach can improve the sustainability of FE cultivation. The proposed execution procedure for this strategy is outlined below:

1. Communal work: The government, through town committees and expansion officers, encourages agriculturists and groups to engage in communal work during various stages of cultivation, such as land preparation or planting. This will diminish the workload for agriculturists with respect to work costs, and communal work can be organized on agricultural land based on farmers' requirements and preferences.
2. Agriculturists orientation: The government allocates extra expansion workers to encourage more gatherings between agriculturists, expansion laborers, and expansion specialists. Extension workers play a crucial role in providing assistance to farmers and promoting a comprehensive understanding of farming practices. Establishing interactive farmers' groups is essential to foster changes and innovation in farming practices for the benefit of the community.
3. Agrarian cultural information: The Humbang Hasundutan area within the Lake Toba Locale offers potential for holding agritourism occasions with social importance. The government routinely organizes events showcasing local cultural values. Annual events celebrating local traditions, such as agrotourism activities during harvest seasons, provide opportunities for visitors to engage with local farmers and select crops. Additionally, regional cultural gatherings contribute to promoting agricultural heritage and practices. (Dinanta et al. 2021; Juhandi et al. 2023).

Third Strategy: Enhancement of All Sectors

Advancement of all sectors, excluding the environment, constitutes the final strategy. The key values of this strategy align with those outlined in strategies one and two, with additional emphasis on institutional perspectives (such as the frequency of extension services, land tenure extension, and resolution of farming conflicts), innovative technological aspects (including innovation appropriation and significance, and farmers' responsiveness to new technology), and promotional perspectives (covering advancements in marketing, knowledge of fair pricing, and consumer relationships).

This strategy serves as an effective approach to improve the sustainability of FE agriculture. Supplementary undertakings tasks within this strategy involve various stakeholders:

1. Government-private sector partnership: The government engages in a partnership with the private sector to provide financial aid, insurance, and marketing assistance to the processing businesses. In addition, the government offers subsidized agricultural inputs, capital, and technology, specifically targeting sustainable agriculture. Village governments and agricultural extension officers will be involved in the transfer of knowledge and technology.
2. Active farmer engagement: Agriculturists/farmers will be encouraged to actively engage in group activities to gain knowledge on sustainable agriculture, technology adoption, communal work, and farming culture, with the close involvement of extension staff.
3. Incentives to private firms: The government provides incentives to private firms, including banks, cooperatives, and processing industries, to support farmers by offering them finanial aid, services, and access to markets.

These three strategies can be viewed as complementary methods that require efficient coordination and collaboration among government agencies, commercial sectors, and farmers. The third strategy, in particular, demonstrates the highest level of sustainability, indicating that enhancing the sustainability of FE farming necessitates development in all elements rather than focusing on just one (Dinanta et al. 2021; Juhandi et al. 2023).

Conclusion

The establishment of FEs in Indonesia dates back to the early planning period in 1955 until and continues to the present day. The purpose of these FEs is to boost agricultural and livestock production with the aim of reducing dependency on imported goods and achieving self-sufficiency. However, numerous FEs have been established without considering variables such as suitable location, water accessibility, climatic conditions appropriate for the crop cultivation, technological innovation, and collaboration between the government and agriculturists. Consequently, the output of these FEs has often fallen short of expectations. Land proprietorship further hinder the sustainability of these projects. Labor concerns, including farmers' skills and capabilities, must be addressed through continuous support and training to ensure effective management of agricultural land and satisfactory agricultural outcomes. Additionally, policies or regulations formulated by the Government play a significant role in determining the success of FEs in Indonesia.

The execution of the FE policy must consider several factors to ensure local market protection, increase agricultural productivity, improve the quality of human resources, and strengthen coordination among stakeholders in agricultural development. Environmental issues such as carbon emissions from deforestation, biodiversity loss, diminished water catchment zones, and flood disasters must be addressed. Food security should be based on ecological agriculture, with commodities selected according to regional soil and climate conditions. Other important considerations include geological conditions, social and cultural characteristics of local communities, reginal development diversity, accessibility limitations. Involving local communities in the planning and implementation process is crucial, as their involvement fosters a sense of ownership and responsibility in maintaining and managing the land properly. Developing an agricultural knowledge system through research and development with the involvement of universities and the private sector can foster innovation and enhance productivity. This colaborative approach will help make the cultivating results more sustainable and improve crop yield.

Disclaimer

None.

References

Adi, Saputra. 2021. [*Challenges of the Food Estate Programme, in Maintaining Food Security*]. Opinion Column of Agromina Bahari Clinic, Faculty of Land, Universitas Gadjah Mada.

Astuti, N. A. R. 2020. *Penjelasan KLHK soal Hutan Lindung Bisa diubah Jadi Food estate*. https://news.detik.com/berita/d-5257973/penjelasan-klhk-soal-hutan-lindung-bisa-diubah-jadi-food-estate. Accessed on: 19th June 2024.

BBC. 2021. Food estate: [*Food estate project in protected forest, environmentalists warn disaster and conflict with indigenous communities 'inevitable'*]. https://www.bbc.com/indonesia/indonesia-54990753.

Badan Penelitian dan Pengembangan Pertanian (Balitbangtan). *Buku Pintar Food estate*. http://www.litbang.pertanian.go.id/infoaktual/819/file/bagian-1, (2011). Accessed on: 19th June 2024

Basundoro, Febrian, A., Sulaeman, F. H. 2020. [Reviewing food estate development as a national resilience strategy in the Covid-19 pandemic era]. *Jurnal Lemhannas RI* 8(2): 27-41.

Dinanta, Galih. P., Afifuddin, Hasanah, F. A., Sadmono, H., Ghifari, M. R. 2021. Geo-Assessment of National Food Estate project at North Sumatra, Indonesia. *IEEE Asia-Pacific Conference on Geoscience, Electronics and Remote Sensing Technology* (AGERS).

Juhandi, Dany, et al. "*Food Estate vs. Non-Food Estate Farming: Seven Aspect to Assess Farm Sustainability in North Sumatra, Indonesia.*" Non-Food Estate Farming: Seven Aspect to Assess Farm Sustainability in North Sumatra, Indonesia, (2023).

Kamil, Sitti Utami Rezkiawaty, and L. O. M. Elwan. "BUNGA RAMPAI FENOMENA SOSIAL POLITIK DIMASA PANDEMI." *Laboratorium Ilmu Komunikasi* (2021).

Kementerian ESDM. *Pasal 1 angka 2 Undang-Undang Nomor 41 Tahun 1999 tentang Kehutanan*, (1999).

Kompas.com. Mengenal Food estate, *Program Pemerintah yang Disebut Dapat Meningkatkan Ketahanan Pangan.* Accesed (Juli 2021). https://www.kompas.com/tren/read/2020/07/09/183800365/mengenal-food-estate-program-pemerintah-yang-disebut-dapat-meningkatkan?page=all.

Lasminingrat, Lula, and Efriza Efriza. "The development of national food estate: The Indonesian food crisis anticipation strategy." *Jurnal Pertahanan dan Bela Negara* 10.3 (2020): 229-248.

Mutia, An Nissa Ayu, and Nadia Astriani. "Pengaturan Pembangunan Food Estate pada Kawasan Hutan untuk Mewujudkan Ketahanan Pangan di Indonesia." *Bina Hukum Lingkungan* 6.2 (2022): 224-240.

Neilson, Jeff, and Josephine Wright. "The state and food security discourses of Indonesia: Feeding the bangsa." *Geographical Research* 55.2 (2017): 131-143.

Nurrochmat, D. R. 2020. [*The Future of Forest Resources from the Perspective of the Job Creation Law*]. paparan pada Diskusi Law Disclosure FH Unpad.

Puspitojati, Triyono, Encep Rachman, and Kirsfianti L. Ginoga. "Hutan Tanaman Pangan: Realitas, Konsep dan Pengembangan." *Kanisius*, (2014).

Puspitojati, Triyono. "Analisis kelayakan pangan sebagai hasil hutan." *Jurnal Analisis Kebijakan Kehutanan* 12.3 (2015): 249-262.

Rasman, Alsafana, Eliza Sinta Theresia, and M. Fadel Aginda. Analisis implementasi program food estate sebagai solusi ketahanan pangan Indonesia. *Holistic: Journal of Tropical Agriculture Sciences* 1.1 (2023).

Rozaki, Zuhud. COVID-19, agriculture, and food security in Indonesia. Reviews in *Agricultural Science* (2020) 8:243-260.

Sianipar, Binsar, and Audrey G. Tangkudung. "Tinjauan ekonomi, politik dan keamanan terhadap pengembangan food estate di Kalimantan Tengah sebagai alternatif menjaga ketahanan pangan di tengah pandemi Covid-19." *Jurnal keamanan nasional* 6.2 (2020): 235-248.

Sianipar, Binsar, and Audrey G. Tangkudung. "Tinjauan ekonomi, politik dan keamanan terhadap pengembangan food estate di Kalimantan Tengah sebagai alternatif menjaga ketahanan pangan di tengah pandemi Covid-19." *Jurnal keamanan nasional* 6.2 (2020): 235-248.

Sigalingging, R., et al. "Analysis of energy used on shallot farming in Food Estate, Hutajulu, North Sumatra." *IOP Conference Series: Earth and Environmental Scienc*e. Vol. 1241. No. 1. IOP Publishing, (2023).

Sigalingging, R., et al. *Life cycle assessment of shallot farming in Food Estate Hutajulu,* North Sumatra, Indonesia. (2023).

Stafford-Smith, Mark, et al. Integration: the key to implementing the Sustainable Development Goals. *Sustainability science* (2017) 12:911-919.

Suharyanto, H. "*Ketahanan Pangan. Sosial Humaniora,* 4 (2), 186–194." (2011).

Vinolina NS. "*Food Estate sebagai Strategi Ketahanan Pangan Nasional Pada Masa Pandemi Covid-19 Pemikiran Guru Besar USU Tentang Corona Virus Disease-19.*" USU Press, (2020).

Virananda, I. Gede Sthitaprajna, and Mohamad Ikhsan. "How COVID-19 affects food security in Indonesia." *Institute for Economic and Social Research* (2021).

Zannati, Anky. "Ketahanan Pangan di Masa Pandemi COVID-19: Langkah Indonesia dengan Food Estate." *BioTrends* 11.2 (2021): 29-34.

Chapter 13

A Synthesis of Graphene Oxide/Bacterial Cellulose Nanocomposites

Saharman Gea*
Tamrin
and Andriayani
Department of Chemistry, Faculty of Mathematics and Natural Sciences, Universitas Sumatera Utara, Medan, Indonesia
Cellulosic and Functional Material Research Center, Universitas Sumatera Utara, Medan, Indonesia

Abstract

Graphene Oxide/Bacterial Cellulose (GO/BC) nanocomposites represent a pioneering class of materials that harness the unique properties of graphene oxide (GO) and bacterial cellulose (BC) to create composites with enhanced mechanical, electrical, and thermal properties. This chapter explores various synthesis methods for GO/BC nanocomposites, including *in-situ* synthesis, *ex-situ* mixing, layer-by-layer assembly, chemical bonding, and novel approaches such as microwave-assisted and enzyme-mediated synthesis. Each method is analyzed for its ability to manipulate the distribution and interaction of GO within the BC matrix, impacting the composite's overall performance. The properties of GO/BC nanocomposites are extensively compared with traditional materials, highlighting their superior mechanical strength, enhanced electrical conductivity, increased thermal stability, and improved barrier properties. Technical challenges such as scalability, uniformity, and environmental impact are discussed

* Corresponding Author's Email: s.gea@usu.ac.id.

In: Natural Resources, Environment, and Food and Energy Security …
Editors: Mohammad Basyuni and Severino Salmo III
ISBN: 979-8-89530-283-5

in detail, alongside potential applications in fields ranging from flexible electronics and biomedical devices to environmental engineering. Future research directions are proposed, focusing on optimizing synthesis processes, enhancing material properties for specific applications, and addressing the sustainability and commercial viability of GO/BC nanocomposites. This comprehensive analysis underscores the transformative potential of GO/BC nanocomposites in advancing material science and engineering, paving the way for innovative solutions in numerous technological arenas.

Keywords: graphene oxide, bacterial cellulose, nanocomposite

Introduction

Bacterial cellulose is synthesized from glucose by bacteria like *Komagataeibacter xylinus* or *Acetobacter xylinum* in a culture medium. BC is well known to have high purity, crystallinity, tensile strength, flexibility, and water holding capacity (Gea, 2010). These attributes make BC a promising material for various applications including medical devices such as drug delivery systems, wound dressings (Pasaribu et al. 2023; Pasaribu et al. 2020), and artificial blood vessels, as well as in food products as a thickener or dietary fiber (Saleh et al. 2024). Despite its impressive mechanical properties, BC's performance is limited in dry conditions where it becomes brittle and loses elasticity. This poses challenges in environments that require durability across both wet and dry states. Additionally, BC's non-conductive nature restricts its applications in electronic components like biosensors and conductive films (Prilepskii et al. 2023; Amaturrahim et al. 2018).

On the other hand, graphene oxide (GO) is a chemically modified form of graphene, obtained from the oxidation of graphite. It stands out due to its exceptional mechanical properties, thermal stability, and electrical conductivity, which are attributed to its atomic-thick layered structure and oxidized surface rich in functional groups such as hydroxyl, epoxy, and carboxyl (Saleh et al. 2024; Prilepskii et al. 2023; Amaturrahim et al. 2018). These functional groups not only render GO hydrophilic but also allow for further chemical modifications, expanding its use across various fields including electronics and nanotechnology. However, the challenges associated with the integration of GO into practical applications often arise due to its tendency to agglomerate, leading to an uneven distribution.

Researchers have explored the introduction of GO into the BC matrix to enhance its properties. This composite shows higher mechanical properties, thermal stability, and electrical conductivity, suitable for certain applications such as electrically conductive scaffolds for tissue regeneration, implantable devices, and flexible electronics. However, evenly distributing GO within BC and maintaining the composite's stability are challenging (Saleh et al. 2024; Prilepskii et al. 2023). The synergy between BC and GO could potentially overcome these limitations, making it a promising candidate for advanced biomaterial applications. Yet, achieving the right balance between tensile strength and porosity to preserve BC's inherent properties while maximizing GO remains a key area of development.

Rationale for Combining GO and Bacterial Cellulose

The integration of GO with BC is driven by the complementarity of their respective properties, which synergistically enhance the functionality of the resultant composites. GO's rigid and highly conductive nature pairs with the flexibility and biocompatibility of BC to create materials that are not only structurally versatile but also functionally diverse. Motivations for combining these materials highlight their potential to achieve practical applications.

The abundant functional groups on GO's surface allow for extensive chemical modifications, which can be tailored into specific functionalities such as antimicrobial properties or increased thermal stability, broadening the scope of BC-based materials in practical applications. The potential for low-cost production and sustainability are significant drivers behind the research into GO/BC composites. Both materials can be produced from abundant or renewable resources, presenting an environmentally friendly.

GO offers transformative enhancements to BC, making it more suitable for a range of high-performance applications. Here are specific areas where GO significantly impacts BC:

1. Mechanical Strength and Toughness

At the molecular level, the interaction between GO and BC involves a complex interaction of hydrogen bonding and van der Waals forces, which significantly enhances the interfacial adhesion, crucial for maintaining the mechanical integrity under stress (Amaturrahim et al. 2018; Luo et al. 2019).

2. Electrical Conductivity

The natural insulator of BC limits its composite phase to be used in electrical and electronic applications. Introducing GO transforms the composite into a conductive material (Prilepskii et al. 2023; Wang et al. 2023) which is particularly useful for developing sensors, conductive films, and components for electronic devices (Purwandari et al. 2018).

3. Thermal Stability

While being sensitive to high temperatures, the addition of GO into BC films enhances the thermal stability of the composite (Amaturrahim et al. 2018; Pinto et al. 2020), making it suitable for applications of heat resistance materials.

4. High-Barrier Properties

The introduction of GO improves the barrier properties of BC especially against gases like oxygen and water vapor. This enhancement is valuable in packaging applications by extending durability against moisture (Khajavi et al. 2020).

5. Chemical Resistance

GO confers additional chemical resistance to the BC matrix by reducing the permeability of chemicals. It also enhances the composite's resistance to acids, bases, and organic solvents.

6. Biocompatibility and Bioactivity

When incorporated into BC, GO can also enhance its biocompatibility and bioactivity (Luo et al. 2019), which are essential for biomedical applications such as tissue engineering scaffolds, wound dressings, and drug delivery systems. GO's surface is functionalized with bioactive molecules, further enhancing its interaction between the material and biological systems.

7. Optical Properties

GO significantly influences the optical properties of BC composites, such as opacity and transparency. By adjusting the GO content and its dispersion of GO within BC, it is possible to tailor the optical characteristics for specific applications, like optoelectronic devices or transparent films (Dellansyah et al. 2022).

Interaction Mechanism at the Molecular Level

Understanding the interaction mechanism at the molecular level between GO and BC in the form of a nanocomposite is a critical aspect that determines the material's properties and performance. These interactions primarily involve noncovalent bonding, which includes hydrogen bonding, van der Waals forces, and π-π interactions.

1. Hydrogen bond formation

Hydrogen bonding plays a significant role in the interaction between GO and BC. BC is rich in hydroxyl (–OH) groups due to its polysaccharide nature, featuring a glucose monomer structure with three hydroxyl groups available for bonding. Meanwhile, GO is chemically exfoliated graphene that includes various oxygen-containing functional groups such as carboxyl (–COOH), hydroxyl (–OH), and epoxy (–O–) groups on its basal planes and edges (Amaturrahim et al. 2018; Urbina et al. 2020).

When GO and BC are mixed, the hydroxyl groups of BC form hydrogen bonds with the oxygen-containing groups on GO. This interaction facilitates the uniform dispersion of GO sheets within the BC matrix, enhancing the mechanical integration and stability of the nanocomposite. The hydrogen bonds effectively "lock" the GO sheets in place within the BC network, preventing agglomeration and ensuring uniform distribution of enhanced properties (like mechanical strength and barrier properties) throughout the material.

2. Van der waals forces

Aside from hydrogen bonding, van der Waals forces also play a role in the interaction between GO and BC (Urbina et al. 2020). These forces are weaker than hydrogen bonds but are significant in contributing to the overall stability of the nanocomposite. Van der Waals forces arise due to the induced polarization from one molecule to another, facilitating attraction over short distances. In the GO/BC composite, these interactions help to further stabilize the structure, maintaining the integrity of the nanocomposite under various conditions.

3. π-π Interactions

The aromatic regions of GO which retain the sp2 hybridized carbon lattice similar to graphene, engage in π-π stacking interactions with the aromatic components of cellulose degradation products or additives that may be present in the BC matrix. Although BC itself does not contain aromatic rings, the processing or modification of BC might introduce aromatic compounds that can interact with GO (Amaturrahim et al. 2018). These π-π interactions are beneficial for enhancing the electrical conductivity of the nanocomposite, which is crucial for applications in electronics and sensing.

4. Impact of nanocomposite synthesis

Noncovalent interactions (hydrogen bonding, van der Waals forces, and π-π interactions) collectively influence the physical properties of the GO/BC nanocomposite. They help to improve the composite's tensile strength, elasticity, thermal stability, and electrical conductivity. Moreover, they ensure that the GO sheets are well integrated into the BC matrix, which is vital for achieving a homogeneous material that performs consistently in its intended applications.

GO/BC Nanocomposite Synthesis

In-Situ Synthesis

In-situ synthesis involves the integration of GO directly into the BC matrix during its biosynthesis. This method is highly effective for fabricating nanocomposites with a strong interfacial bond between GO and BC, leading to enhanced mechanical and thermal properties of the resulting material.

In this process, GO is dispersed into the culture medium used for growing bacteria such as *Komagataeibacter xylinus* or *Acetobacter xylinum.* The GO sheets become entangled within the nascent cellulose fibers as they are produced by the bacteria. This entanglement occurs at the molecular level, ensuring that GO is uniformly distributed within the BC matrix. The advantage of this method is the high level of integration and interaction between the GO sheets and cellulose fibers, which can significantly enhance the physical properties of the composite. For instance, Gabrys et al. (2023) reported significant increase in tensile strength 150% higher of GO/BC nanocomposite than that of pure BC which synthesized *in-situ* and reported.

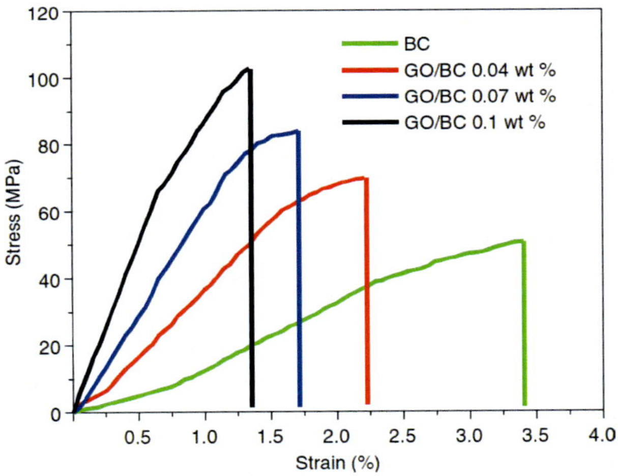

Figure 1. Stress-strain curves of grapheme oxide/bacterial cellulose (GO/BC) [6].

One notable study by Amaturrahim et al. (2018) demonstrated that GO can be effectively dispersed within the network of BC using an agitation method. The analysis revealed a broadened peak at 3371 cm^{-1}, along with changes in the relative intensity of the C-O stretching vibrations in the GO/BC nanocomposites, suggesting that the hydrogen bonds in BC were altered. This disruption is indicative of potential effective interactions, such as hydrogen bonding, between the oxygen-containing groups on GO and the hydroxyl groups in BC. These interactions are further evidenced by the observed decrease in the absorption of stretching vibrations associated with C-O epoxy, C-O alkoxy, and bending of OH groups. The incorporation of GO not only enhanced the thermal and mechanical properties (Figure 1) of the nanocomposite but also resulted in larger inter-fibril pores and a reduction in the size of the fibrils themselves (Amaturrahim et al. 2018). This highlights the dual impact of GO integration on the structural and functional attributes of BC.

The primary advantage of this method is the potential for achieving a homogenous distribution of GO within the BC matrix, which is critical for applications requiring uniform mechanical properties and conductivity. However, the main challenge lies in controlling the dispersion of GO and maintaining the viability and productivity of the bacteria during synthesis. Overcoming these challenges requires careful optimization of GO concentration and culture conditions to ensure that the bacterial activity is not adversely affected.

Ex-Situ **Mixing**

Ex-situ synthesis involves combining pre-synthesized GO with BC after both components have been independently produced. This method allows for greater control over the composite's final properties, especially in terms of the distribution and concentration of GO within the BC matrix. The main techniques used in *ex-situ* synthesis include blending, casting, and freeze-drying.

The main advantage of *ex-situ* synthesis is the flexibility it offers in terms of adjusting the properties of the composite by manipulating the concentration and distribution of GO. This method is also beneficial for applications requiring specific thicknesses or shapes of the composite material, as it can be easily cast or molded.

However, one of the significant challenges with *ex-situ* synthesis is ensuring a uniform distribution of GO within the BC. Inadequate mixing can lead to clumping or uneven distribution, which can adversely affect the mechanical and electrical properties of the final product. Additionally, the physical bonding achieved through *ex-situ* methods is often weaker than the molecular-level integration seen in *in-situ* synthesis, which can result in composites with less optimal performance characteristics.

Layer-by-Layer Assembly of GO/BC Nanocomposite

Layer-by-Layer (LbL) Assembly is a precise method for fabricating nanocomposites by alternately depositing layers of GO and BC to build up a composite structure with highly controlled thickness and composition. This method allows for the meticulous manipulation of the nanostructure, resulting in composites with tailored properties.

A layer-by-layer assembly done by (Luo et al. 2018), grown the GO/BC films on a base membrane, with each cycle involving the spraying of a GO-containing culture medium onto the base membrane, followed by biosynthesis to grow the hydrogel, leads to greatly enhanced mechanical properties compared to those of bare BC.

The primary advantage of the LbL assembly method is its ability to precisely control the microstructure of the composite, which significantly influences its physical and chemical properties. This precise control allows for the optimization of the composite for specific functionalities, such as increased electrical conductivity, improved mechanical properties, or enhanced thermal resistance.

However, the LbL assembly method comes with its own set of challenges. It is typically more time-consuming and labor-intensive than other methods, making it less suitable for large-scale production. The need for multiple processing steps (coating, drying, curing) for each layer also introduces variability in the quality of the final product if not meticulously controlled. Furthermore, achieving uniform layer thickness across a large area can be challenging, which is crucial for maintaining consistent properties throughout the material.

Chemical Bonding Methods for GO/BC Nanocomposites

Chemical Bonding Methods involve the use of chemical agents to create covalent or ionic bonds between GO and BC. This approach enhances the interfacial adhesion between the two materials, leading to composites with improved mechanical, thermal, and electrical properties.

The main advantage of using chemical bonding methods in the fabrication of GO/BC composites is the strong interfacial adhesion they promote. This bonding significantly enhances the durability and functional stability of the composite under various physical and environmental conditions. Additionally, chemical bonding can introduce new functionalities to the composite, such as increased hydrophobicity or flame retardancy, depending on the chemicals used for bonding.

However, there are several challenges associated with chemical bonding methods. The chemical agents used may require specific conditions, such as elevated temperatures or specific pH levels, to react effectively. This can complicate the manufacturing process. Moreover, some chemical agents might be toxic or environmentally harmful, posing risks during handling and disposal. The complexity of the chemical processes involved can also increase the cost of production, making it less economically viable for large-scale applications.

Current Challenges and Discussion

Technical Challenges and Limitations

While the synthesis of GO/BC nanocomposites presents promising advancements in material science, several technical challenges and limitations must be addressed to fully harness their potential.

Uniformity and Distribution of GO

One of the primary challenges in synthesizing GO/BC nanocomposites is achieving a uniform distribution of GO within the BC matrix. GO has a tendency to agglomerate due to van der Waals forces and π-π interactions among its sheets, which can lead to inconsistent material properties. Techniques like ultrasonication and high-shear mixing are often employed to disperse GO effectively, but these methods require precise control and can be difficult to scale up while maintaining uniformity.

Novel Synthesis Approaches

There is potential for the development of innovative nanoparticle synthesis methods in the production of GO/BC nanocomposites. These approaches aim to optimize and enhance the functional integration of GO and BC, moving beyond conventional techniques to achieve superior material properties and functionalities. Examples include microwave-assisted synthesis, enzyme-mediated reactions for binding GO to BC, electrospinning, supercritical fluid technology, and ultrasonication-assisted synthesis, among others.

Scalability of Production

Scaling the production of GO/BC nanocomposites from laboratory to industrial scales presents significant challenges. Replicating the precise environmental conditions, such as temperature, pH, and humidity, which are crucial for *in-situ* synthesis methods, can be challenging on a larger scale. Additionally, maintaining the quality and consistency of the nanocomposites during large-scale production requires advanced manufacturing processes that can handle the delicate nature of BC and the reactive properties of GO.

Environmental and Health Concerns

Although both GO and BC are considered relatively environmentally friendly, the production and processing of GO can involve chemicals that pose environmental and health risks. For instance, the reduction of GO to reduced GO (rGO), which is often performed to improve conductivity, typically involves chemical reducers that can be toxic. Safe handling practices and disposal methods need to be developed to mitigate these risks.

For instance, the use of strong oxidizing agents in the synthesis of GO and the potential residues left on GO can pose environmental risks if not adequately treated before disposal. Researchers and industry must develop and adhere to protocols for safely handling and disposing of these chemicals, and efforts should be made to replace toxic chemicals with greener alternatives where possible.

Cost-Effectiveness

While BC can be produced cost-effectively, the synthesis and processing of GO, especially when aiming for high-quality and functionalized forms, can be costly. The overall cost-effectiveness of GO/BC nanocomposites can be a limiting factor, especially in applications where cheaper alternatives are available.

Performance Stability

The long-term stability of GO/BC nanocomposites under real-world environmental conditions is another concern. Factors such as UV exposure, moisture, and mechanical wear can degrade the properties of the nanocomposites over time. Ensuring the durability and stability of these materials in various applications remains a significant challenge.

Future Research Directions

The study and development of GO/BC nanocomposites have opened up a variety of possibilities and highlighted several areas where further research could significantly enhance the understanding and application of these materials. Future research directions are crucial for advancing the field and addressing the current limitations and challenges.

Optimization of Synthesis Methods

Continued research is needed to optimize the synthesis methods for GO/BC nanocomposites. This includes refining *in-situ* and *ex-situ* methods to enhance the uniformity and distribution of GO within the BC matrix. Innovations in synthesis could involve exploring new solvents, green chemistry approaches, or advanced mixing and deposition techniques that reduce environmental impact and improve material performance.

Tailoring Material Properties

Further studies should focus on tailoring the properties of GO/BC composites for specific applications. This could involve manipulating the size, functionalization, and concentration of GO, as well as the porosity and alignment of BC fibers. Advanced computational modeling and simulation could also play a role in predicting the outcomes of various composite configurations before physical testing.

Scale-Up and Commercialization

Scaling up the production of GO/BC composites from laboratory to industrial scales remains a significant challenge. Future research should focus on developing scalable production processes that are cost-effective and environmentally friendly. Additionally, studies on the commercialization potential of these composites will be important for moving from research and development into actual market applications.

Interdisciplinary Collaborations

Encouraging interdisciplinary collaborations between material scientists, engineers, biologists, and industry professionals will be crucial for addressing the complex challenges associated with GO/BC nanocomposites. Such collaborations can foster innovation, from new synthesis techniques to novel applications in biomedicine and electronics.

Conclusion

The exploration and development of GO/BC nanocomposites have opened up exciting possibilities across various fields. However, while the potential applications are vast, there are several challenges and future perspectives that need to be addressed to fully realize their capabilities. By continuing to tackle the technical challenges and focusing on targeted research efforts, the future of these materials looks promising. Their continued development will likely provide sustainable and efficient solutions to some of the most pressing challenges in healthcare, environmental engineering, and beyond.

Disclaimer

None.

References

Amaturrahim, S. A., Gea, S., Nasution, D. Y., Hutapea, Y. A. 2018. Preparation of graphene oxide/bacterial cellulose nanocomposite via *in-situ* process in agitated culture. *Asian J. Chem* 30(7): 1564–1568, 2018, doi: 10.14233/ajchem.2018.21244.

Dellansyah, D., Gea, S., Andriayani, Harahap, M., Nainggolan, G. 2022. *Optical Study of Graphene Quantum Dots from Sawahlunto Coal Graphite* 584–586. doi: 10.5220/0010614000002775.

Gabryś, T., Fryczkowska, B., Jančič, U., Trček, J., Gorgieva, S. 2023. GO-Enabled Bacterial Cellulose Membranes by Multistep, *In-situ* Loading: Effect of Bacterial Strain and Loading Pattern on Nanocomposite Properties. *Materials (Basel).* 16(3). doi: 10.3390/ma16031296.

Gea, S. 2010. Innovative Bio-nanocomposites Based on Bacterial Cellulose. A thesis submitted to the Queen Mary, University of London

Khajavi, M. Z., Ebrahimi, A., Yousefi, M., Ahmadi, S., Farhoodi, M., Alizadeh, A. M., Taslikh, M. 2020. Strategies for Producing Improved Oxygen Barrier Materials Appropriate for the Food Packaging Sector. *Food Eng. Rev.* 12(3): 346–363. doi: 10.1007/s12393-020-09235-y.

Luo, H., Ao, H., Peng, M., Yao, F., Yang, Z., Wan, Y. 2019. Effect of highly dispersed graphene and graphene oxide in 3D nanofibrous bacterial cellulose scaffold on cell responses: A comparative study. *Mater. Chem. Phys* 235(3): 121774, 2019, doi: 10.1016/j.matchemphys.2019.121774.

Luo, H., Dong, J., Yao, F., Yang, Z., Li, W., Wang, J., Xu, Z., Hu, Y. 2018. Layer-by-Layer Assembled Bacterial Cellulose/Graphene Oxide Hydrogels with Extremely

Enhanced Mechanical Properties. *Nano-Micro Lett.* 10(3): 1–10. doi: 10.1007/s40820-018-0195-3.

Pasaribu, K. M., Ilyas, S., Tamrin, Radecka, I., Swingler, S., Gupta, A., Stamboulis, A. G., Gea, S. 2023. Bioactive bacterial cellulose wound dressings for burns with collagen in-situ and chitosan ex-situ impregnation. *Int. J. Biol. Macromol.* 230(1): 123118, 2023, doi: 10.1016/j.ijbiomac.2022.123118.

Pasaribu, K. M., Gea, S., Ilyas, S., Tamrin, Radecka, I. 2020. Characterization of bacterial cellulose-based wound dressing in different order impregnation of chitosan and collagen. *Biomolecules* 10(11): 1–15, 2020, doi: 10.3390/biom10111511.

Pinto, S. C., Golcalves, G., Sandoval, S., Lopez-Periago, A. M., Borras, A., Domingo, C., Tobias, G., Duarte, I., Vicente, R., Marques, P. A. A. 2020. Bacterial cellulose/graphene oxide aerogels with enhanced dimensional and thermal stability. *Carbohydr. Polym* 230(2): 115598, 2020, doi: 10.1016/j.carbpol.2019.115598.

Purwandari, V., Gea, S., Wirjosentono, B., Haryono, A. 2018. Synthesis of graphene oxide from the Sawahlunto-Sijunjung coal via modified hummers method. *AIP Conf. Proc* 2049(12). doi: 10.1063/1.5082470.

Prilepskii, A., Nikolaev, V., Klaving, A. 2023. Conductive bacterial cellulose: From drug delivery to flexible electronics. *Carbohydr. Polym.*, 313(12): 120850, 2023, doi: 10.1016/j.carbpol.2023.120850.

Saleh, A. K., Ray, J. B., El-Sayed, M. H., Alalawy, A. I., Omer, N., Abdelaziz, M. A., Abouzeid, R. 2024. Functionalization of bacterial cellulose: Exploring diverse applications and biomedical innovations: A review. *Int. J. Biol. Macromol* 264(P1): 130454, 2024, doi: 10.1016/j.ijbiomac.2024.130454.

Urbina, L., Eceiza, A., Gabilondo, N., Corcuera, M. A., Retegi, A. 2020. Tailoring the *in-situ* conformation of bacterial cellulose-graphene oxide spherical nanocarriers. *Int. J. Biol. Macromol.* 63: 1249–1260. doi: 10.1016/j.ijbiomac.2020.07.077.

Wang, C., Wu, R., Wang, L., Wang, X. 2023. Multifunctional flexible graphene oxide/bacterial cellulose composite paper platforms for real time monitoring sweat and strain in wearable devices. *Chem. Eng. J.* 481(10): 148390. doi: 10.1016/j.cej. 2023.148390.

Chapter 14

The Development of a Resilient Food System in North Sumatra

Elisa Julianti[1,*]
Diana Sofia Hanafiah[2]
and Yaya Hasanah[2]
[1]Department of Food Science, Faculty of Agriculture,
Universitas Sumatera Utara, Medan, Indonesia
[2]Department of Agrotechnology, Faculty of Agriculture,
Universitas Sumatera Utara, Medan, Indonesia

Abstract

A resilient food system is defined by its long-term ability to ensure sufficient, appropriate, and accessible food for all individuals at various levels, even in the face of unexpected disruptions. The food system is influenced by a variety of interacting factors that affect food stability and nutritional security, including environment, human resources, inputs, processes, infrastructure, institutions, and activities related to food production, processing, post-harvest management, distribution, preparation, and consumption. The objective of this review is to summarize the concept of developing a resilient food system, particularly in North Sumatra. Information was collected from the databases of the Ministry of Agriculture of the Republic of Indonesia, the Central Statistics Agency, and research repositories at the Universitas Sumatera Utara, as well as from global databases such as Google Scholar, Scopus, and Web of Science using keywords "resilient," "food system," "germ plasm," and additional keywords

* Corresponding Author's Email: elisa1@usu.ac.id.

In: Natural Resources, Environment, and Food and Energy Security …
Editors: Mohammad Basyuni and Severino Salmo III
ISBN: 979-8-89530-283-5

"rice," "corn," "soybean," "sweet potato," "cassava," "shallot," "garlic," and "chili." The discussion includes the utilization of germplasm commodities, climate suitability, and considers aspects of local plant conservation strategies, cultivation technology, and post-harvest technology to improve the productivity of germplasm commodities in North Sumatra, particularly rice, corn, soybeans, chili, shallots, garlic, sweet potatoes, and cassava. Preserving plant germplasm, especially the diversity of endemic plants in North Sumatra, can be done through ex situ, in situ, and on-farm conservation methods.

Keywords: climate, food system, self-sufficiency, technology, topography

Introduction

Food is the primary fundamental requirement of every human being; therefore, fulfilling food requirements is a human right and plays a significant role in the existence of a nation. Food system resilience is a critical issue for ensuring the wellbeing of societies worldwide. A compromise in food security can lead to political and social unrest, endangering economic and national stability.

The current concept of food systems implies that these systems are influenced by various interacting factors that support food stability and nutritional security. Food system models and policies should not just focus on the production or individual components of the food system (Reardon and Timmer, 2012), but also adopt a holistic approach to achieve stable food, nutrition, and safety. A food system encompasses all elements (environment, humans, inputs, processes, infrastructure, institutions, etc.) and activities related to the production, processing, post-harvest management, distribution, preparation, and consumption of food, as well as the outcomes of these activities, including socioeconomic and environmental outcomes (HLPE, 2017; FAO et al. 2018).

According to Tendall et al. (2015), there are three accessible entry points for the comprehensive development of food system resilience: 1) national or regional food systems, which encompass all value chains that contribute to food security in a specific area; 2) individual food value chains of agricultural commodities, which range from local to global levels and form national and regional food systems; and 3) the individual perspective within the value chain, including smallholder livelihoods, household food security, and consumer health, among others, as components of the food system.

The province of North Sumatra in Indonesia has a total area of approximately 181,860.65 km^2, which includes 72,981.23 km^2 of land and 108,878.77 km^2 of sea. Administratively, North Sumatra Province is divided into 25 districts and eight cities, 444 sub-districts, 5,417 villages, and 693 urban neighborhoods (North Sumatra Province, 2021). In 2023, the population of North Sumatra Province was 15, 386, 649 (Statistics of Sumatera Utara, 2024). Like other Indonesian provinces, North Sumatra has advantages in terms of its natural resources, including agricultural commodities, plantations, fisheries, livestock, and biodiversity. However, ensuring food security for its population remains a major challenge dua to a variety of factors, such as natural disasters, climate change, pandemics, supply chain issues, rising food prices, infrastructure challenges, conflicts, and a decline in the agricultural labor force. Genetic resources or germplasm have a crucial role in supporting food sovereignty, nutritional adequacy, and increasing competitiveness in the global market. The importance of genetic resources in agriculture and the threats to their sustainability have become the subject of discussion at the international level. In addition, the need to protect agricultural genetic resources and achieve the Sustainable Development Goals to achieve food security and sustainable agriculture has increased.

Agricultural genetic resources are a key element for achieving and maintaining food security. Together with land and water, crop genetic resources is a key component underlying food and agricultural security (FAO, 2014). The genetic resources of Food and Agricultural Crops includes modern crops, breeding lines, genetic stocks, old crops, ecotypes, varieties, wild relatives and their wild species (FAO, 2016). According to FAO (1997), around 80,000 plant species can be used as a food source, but only around 7,000 species can be cultivated as agricultural crops, and currently only 30 species contribute 90% of the world's food needs. The six main species, such as rice, corn, wheat, potatoes, barley and soybeans, produce greater production than all other crops.

This chapter discusses the concept of establishing a resilient food system in North Sumatra, elucidating the potential of germplasm, climate suitability, and cultivation technology to increase the productivity of germplasm commodities, as well as the contribution of post-harvest technology to support food system resilience in North Sumatra. Information was collected from the databases of the Ministry of Agriculture of the Republic of

Indonesia, the Central Statistics Agency, and research repositories at the Universitas Sumatera Utara, as well as from global databases such as Google Scholar, Scopus, and Web of Science using keywords "resilient," "food system," "germ plasm," and additional keywords "rice," "corn," "soybean," "sweet potato," "cassava," "shallot," "garlic," and "chili." The discussion includes the utilization of germplasm commodities, climate suitability, and considers aspects of local plant conservation strategies, cultivation technology, and post-harvest technology to improve the productivity of germplasm commodities in North Sumatra, particularly rice, corn, soybeans, chili, shallots, garlic, sweet potatoes, and cassava.

Germplasm Diversity in North Sumatra That Supports Food System Resilience

Plant genetic resources are part of the germplasm and play an important role in supporting plant breeding programs. Currently, global climate change is a major issue with the potential to threaten the availability of plant genetic resources, by means of global warming and natural disasters. The use of superior varieties is an essential component of efforts to fulfill national food demands.

The increasing use of new varieties by farmers has driven out local varieties (landraces) of which some are now extinct. Therefore, the existing germplasm must be preserved to guarantee the availability of genes both today and in the future. Genes that do not appear to be useful may be required in the future to create new superior varieties. Therefore, plant genetic resources must be conserved so that they are sustainably available to support food availability and security.

Food plant breeding programs that produce new superior varieties with high productivity and yield stability require sources of genes for plant traits that facilitate the creation of new superior varieties. Desirable characteristics include a high yield potential, improved adaptability to suboptimal environmental conditions, resistance to major pests and diseases, a shorter life cycle, and better nutritional content and quality.

North Sumatra has local plants that are well adapted to its agricultural environment and have been cultivated for generations by the community. The local food and horticultural plants in North Sumatra have distinctive and specific morphological and quality characteristics. Initially, plants that were introduced from outside Indonesia were used as food and horticultural plants in Indonesia. Most of these plants have adapted to Indonesia's environment and climate. Environmental conditions, such as altitude, air humidity, soil type, and cultivation technology applied over the years, have influenced the changes in morphological characteristics of the adapted plants. The germplasm diversity related to the food system resilience in North Sumatra described in this chapter are rice, corn, soybean, sweet potato, cassava, chili and garlic

Rice (*Oryza sativa L.*)

North Sumatra has exotic sources of diversity and high economic value. One of the germplasms found in this area is rice, specifically lowland and dryland rice. In North Sumatra Province, lowland rice productivity remains below the national average production of 4.7 tons/ha, whereas dryland rice remains undeveloped and is at risk of extinction. At present and in the future, the contribution of dryland rice is highly anticipated. This is due to the availability of sufficient non-irrigated land and climate change conditions, which tend to be dominant during the dry season. Local cultivars are a strategic source of genetic diversity with great potential for development. The economic value of national food security programs can be enhanced by correcting the deficit characteristics of local cultivars.

If managed properly, local rice can be a valuable genetic asset. Local rice is a primitive rice or cultivar that has been developed over the course of years or even centuries and is influenced by both natural and artificial migration and selection. Local rice has certain advantages, including its adaptability to a variety of climatic conditions and specific lands. Judging from farmers' interests, local rice is easy to obtain and requires minimal maintenance. Additionally, the yield is consistent, the input is low, and the small, slender shape of the grain is favored by farmers and consumers. Local rice has a unique taste and aroma that is appreciated by consumers.

Local rice, both lowland and dryland, typically has a long harvest life of 6 to 7 months. Farmers usually prefer rice with a short harvest time (e.g., 100 d prior to harvest). Additionally, local rice is not resistant to pests and disease attacks, and some of it is susceptible to falling. Another disadvantage is the relatively low productivity, ranging from 1.5 to 2.0 tonnes per hectare (Chaniago, 2017).

There are several types of local rice, including lowland rice, such as Sibosur, Secondong, Bintang Alas, Padang, Siangkat, Siramos, Silumut, and Sipongkol. Dryland rice varieties include Sigambiri, Siodang, Simakattan, Siaruppang, Sikehkeh, Sipala, and Siburu Jabi. Farmers rarely planted certain local rice varieties. Moreover, some of these varieties could no longer be found. The reason for the increasing loss of local rice varieties is that many farmers are reluctant to plant them because of the longer harvest time compared to commercial rice varieties that are widely available in the market, such as Ciherang and Mekongga (Chaniago, 2017). From an economic perspective, planting local rice is more profitable because the seed costs are lower. Agriculturists have extensively used and continue to use several local rice plant varieties to produce rice.

Corn (*Zea mays L.*)

The corn plant is an introduced crop, not native to Indonesia, but since its genetic composition varies dynamically, the population will have a high genetic diversity. The heterozygous genetic composition is heterogeneous and plastic in adapting to different environments.

Corn is highly diverse because its cultivation area is very wide, including both commercial varieties and local cultivars. Most farmers plant corn on dry land during the rainy season, while a small number plant corn on paddy fields after rice. Corn production in North Sumatra is spread across several districts, including Karo, Dairi, and Simalungun, which achieved the highest corn production in 2022 in North Sumatra (Statistics of Sumatera Utara, 2022[a]). Many superior varieties of corn have been released and spread widely in Indonesia. The increasing use of new varieties by farmers has driven out the local varieties (landraces), and some have disappeared.

Germplasm plays an important role as a source of genes, which are basic elements for creating superior varieties. The types of corn produced in Indonesia include hybrid, transgenic, and composite corn. Hybrid corn is the first offspring of a cross between two parents with superior characteristics and attributes. Crossing two homozygous parents produces vigorous F1. Transgenic corn is a type of corn resulting from the insertion of genes such as disease resistance genes, pest resistance genes, or chemical compound resistance genes originating from living or non-living creatures to create a super plant. Composite corn, usually called local corn, is a variety grown in ancient times by local farmers that self-pollinated without human help. The collection of local plant genetic resources plays an important role in supporting future agricultural success. Local corn cultivars are a source of genetic diversity in the development of free-pollinating and hybrid varieties (Ekawati, 2010).

Utilizing technology that is appropriate for the local conditions is another way to increase corn production. Local corn has not been altered technologically, farmers plant corn with seeds that they produce and plant repeatedly. Therefore, the use of superior local seeds, especially in North Sumatra, which have the potential for high yields, needs to be introduced to farmers. Similarly, empowering seed breeder farmers to provide superior local seeds can strengthen efforts to increase corn production. It is anticipated that local corn cultivars will become extinct in the future if they are not conserved. Corn production can be increased through strategies including seed breeding of local varieties that have high production potential to support increased production and business opportunities for farmers, in addition to improving cultivation techniques and using technology in plant cultivation (Darwis, 2018).

Soybean (*Glycine max (L.) Merr.*)

Soybean, an important food crop commodity, occupies a strategic position in North Sumatra and is important as a food crop commodity. Soybeans stand third among the main food commodities after rice and corn. Soybeans are rich in protein content; therefore, this commodity has various uses, especially as a raw material for the food industry and as an ingredient for the animal feed industry.

Among the legume commodities, soybeans have long been known as a source of vegetable protein to the people of Indonesia. Soybeans contain the highest amount of protein among any grain or legume and have the highest amount of carbohydrates, fiber, vitamins, minerals, and phytochemicals that are useful for the prevention and treatment of various chronic diseases.

To date, local farmers have been able to meet domestic needs for soybeans by only 60%. Low local soybean production results from the unwillingness of farmers to grow soybeans because of low profitability. The price of local soybeans is very low because it follows the price of imported soybeans. However, this is an opportunity for Indonesia to capitalize on rising international market prices to develop indigenous soybeans, making prices more competitive. Land is another challenge that domestic soybean commodities confront as they continue to depend on imports. Despite the government's addition of new land for soybean crops, it has remained unutilized.

The districts of Simalungun, Serdang Begadai, Langkat, and Toba were the highest soybean-producing areas in North Sumatra from 2020 to 2022. Approximately 114 superior soybean varieties are certified by the government. The Ministry of Agriculture Republic Indonesia has noted that, so far, there are six national superior varieties of local soybeans that are generally planted: Detaps 1, Dega 1, Dena 1, Dering 1, Anjasmoro, and Grobogan. Each soybean variety has unique characteristics and advantages. North Sumatra farmers generally plant the Anjasmoro variety because it has a large seed size and can grow and adapt well to the North Sumatra area.

Sweet Potato (*Ipomea batatas L.*)

As an alternative food source, sweet potatoes have the potential to serve as carbohydrate source for both feed and industrial applications. North Sumatra Province is a sweet potato production center with various local genotypes and high production potential. This potential needs to be explored and conserve the obtained accessions so that they can be used as genetic material for developing new, superior varieties with high productivity. According to the 2022 Central Statistics Agency data, Dairi, Karo, Samosir, and North Tapanuli are the four districts in North Sumatra with the highest sweet potato production (Statistics of Sumatera Utara, 2022[b]).

Local sweet potato genotypes generally exhibit location-specific characteristics. Therefore, when planted in other areas, production is not optimal. Widely adapted genotypes have the advantage of producing high yields in diverse agroecosystems. Another challenge in sweet potato cultivation is the inconsistent tuber formation on each plant, because each individual yields different results. This variation is related to genetic factors. In sweet potato plants, genetic mechanisms control the formation of storage roots in the form of tubers.

The morphological and production data in the lowlands and highlands of North Sumatra were analyzed in relation to the area of origin of the observations. Two types of sweet potatoes were obtained from the Deli Serdang Regency (lowlands): the local Binjai genotype and the local Hamparan Perak genotype. Two types of sweet potatoes were obtained from the Tanah Karo Regency (highlands): the local genotypes Berastagi and

Simpang Empat. Various genes from each variety were visualized in various characters, illustrating the interaction between genetics and the environment.

Cassava (*Manihot esculenta*)

Cassava is used as a food, feed, and raw material in various industries. Therefore, cassava varieties must be selected according to their intended use. In areas where cassava is consumed directly as food, the varieties that taste good, are soft, and have a low HCN content, are chosen. Based on the HCN content, cassava is divided into sweet (not bitter) cassava, with an HCN content < 40 mg/kg of fresh tubers, and bitter cassava with an HCN content ≥ 50 mg/kg of fresh tubers. High HCN content can cause poisoning in humans and animals; therefore, fresh consumption of cassava is not recommended.

Farmers and other users choose and approve cassava varieties based on the age of the plant, its performance, and its resistance to pests and plant diseases. In general, farmers are extremely passionate about old and local superior varieties that are well known in the larger community, ensuring a clear market.

Serdang Bedagai, Simalungun, North Tapanuli, and Toba were the largest cassava-producing districts in North Sumatra in 2020–2022 with production in the range of 17,528–683,727 tons (Statistics of Sumatera Utara, 2022[c]). Based on the results of morphological and production identification, ten cassava genotypes were identified with different local names depending on the area where the plant was found such as Klanting Merah, Malaysia, Genderuwo Hitam, Roti, Keling, Sawit Gading, Genderuwo Merah, Lampung, Simerah, and Klanting Putih in the Serdang Bedagai Regency. Several of the same local cultivars have also been found in other districts in North Sumatra; however, a limited number have not yet been identified morphologically.

Chili (*Capsicum annum L.*)

The growth of horticulture agribusiness, especially vegetables, aligns with the objectives of agricultural development in Indonesia, including North Sumatra Province, which aims to improve the standard of living for farmers by increasing production and income levels. Chili is one vegetable commodity that is a priority for encouraging agricultural growth in North Sumatra. Chili is an important vegetable commodity with high economic value. This commodity is also a source of income and employment opportunities, making a significant contribution to regional economic development.

Chili is an interesting horticultural commodity for research, in terms of both cultivation and marketing. Chili products cultivated in North Sumatra include large types of chilies, curly red chilies, and cayenne peppers. North Sumatra Province is one of the areas that is suitable for chili cultivation. Chili production in 2021 reached 2.8 million quintals, with a harvest area of 23,240 ha. North Sumatra Province contributed 10.23 percent to the total chili production in Indonesia in 2021. This is one of the important reasons for the development of chili commodities in North Sumatra Province. The chili production centers in North Sumatra are located in several districts. Simalungun Regency contributed 43.43% of the total chili production in North Sumatra, Karo Regency 28.02%, Dairi Regency 8.59%, and Batubara Regency 5.45%. The four largest chili-producing districts/cities in North Sumatra contributed 85.49%, and the remaining 14.51% was contributed by other districts/cities.

In the North Sumatra area, especially in the Karo Regency, many local red chilies are cultivated, which have high productivity, good quality, and long life, and are liked by consumers. The local chilies cultivated by farmers are also diverse, with variations in stem color, canopy width, plant height, and branching density. Currently, farmers in Karo Regency mostly cultivate local chilies with purple, green-purple, and green stems. Batu Bara Regency, Limapuluh Pesisir District in North Sumatra, also has local chili cultivars, namely green, white, and black chili cultivars. The name of this cultivar is based on the color of the young fruits. Farmers continue to grow this local cultivar because the seeds are easy to obtain and can be planted repeatedly, unlike hybrid chili seeds.

Shallots (*Allium ascalonicum L.*)

In North Sumatra Province, Samosir Regency is long known for its local shallots and has become a national shallot-producing area. The Samosir region and its surroundings, including the Samosir, Humbang Hasundutan, Simalungun, and North Tapanuli districts, are the main shallot-producing areas in North Sumatra with the local variety Samosir. Local Samosir shallots have a redder color, lower water content, a spicier taste, a very sharp aroma, and a smaller bulb size than the other shallot varieties.

During 2000-2005, shallot production in Samosir Regency and nearby areas suffered due to a pest infestation that nearly destroyed the entire crop, leading to several years of crop failure. This distress forced farmers to abandon the local variety in favor of Java-origin varieties like the Bali variety. Additionally, difficulties in obtaining seeds contributed to the decline of local Samosir shallots.

Local shallots have a high market price. During the harvest season, shallot farmers can also enjoy the profits from selling shallots, which has become an economic resource for farmers in Samosir. Considering the advantages of the local Samosir shallots, it is necessary to explore shallots that are truly local to Samosir surrounding Lake Toba. Such studies are expected to have a significant impact on the availability of shallot germplasm, its use in breeding programs, and its conservation. In general, farmers in North Sumatra cultivate several types of shallot varieties, including Bima Brebes, Fauji, Super Philip, and Tajuk for highland areas, whereas the varieties Batu Ijo, Maja Cipanas, and Bali Karet are cultivated in lowland areas.

Garlic (*Allium sativum L.*)

Garlic is typically a sterile species that can only be propagated vegetatively by planting tubers. However, several garlic clones capable of producing seeds have recently been collected from the Tien Shan mountainous region of Kyrgyzstan and western China. These seed-producing garlic varieties are essential for plant breeders to increase clonal diversity and develop new superior clones desired by consumers (Currah, 2022).

Both local and introduced garlic varieties, which are generally propagated vegetatively using tubers, have been developed in Indonesia. The development of local varieties is prioritized because they are usually location-specific and have distinct advantages. The garlic clones used in Indonesia were Lumbu Hijau, Lumbu Kuning, and Lumbu Putih. Lumbu Hijau and Lumbu Kuning are appropriate for highland planting, whereas Lumbu Putih is more suitable for planting in the lowlands. Several other local cultivars have considerable potential, including Tawangmangu Baru and Sangga Sembalun, as well as many more local clones that may not have been evaluated.

Local garlic is becoming increasingly difficult to obtain, as farmers are reluctant to grow it due to its inferiority compared to imported garlic from China, which has better quality bulbs and is less expensive. However, the taste quality of local garlic is equally important. Local garlic has a much stronger odor than imported varieties, likely due to its high concentrations of aliin and organosulfur compounds. This strong smell suggests that local garlic could be valuable as a pharmacological ingredient.

The local garlic cultivar Doulu is found in northern Sumatra. This cultivar was obtained from Karo. The results of field research showed that the local garlic cultivar Doulu was distributed only in the Karo and Dairi districts. The availability of Doulu cultivars is rare because people prefer to consume imported garlic. The eruption of Mount Sinabung damaged the cultivation sites of the Doulu cultivar. The imported garlic cloves are physically larger than the local Doulu garlic. The imported garlic had an average length of 3.16 cm, a diameter of 1.59 cm and weighed 5.79 g /clove. Local garlic cloves have an average length of 2.16 cm, a diameter of 0.94 cm and weigh 1.86 g/clove. Based on 2021 Central Statistics Agency data, the largest number of garlic plantations in North Sumatra are in the Humbang Hasundutan and Simalungun Regencies (Statistics of Sumatera Utara, 2021).

Local Plant Conservation Strategies

It is necessary to preserve the plant germplasm, especially the diversity of endemic plants in North Sumatra. Plant germplasm, especially wild and local plants, can be preserved in gene banks, known as ex situ conservation. This plays a crucial role in preventing the extinction of wild and local plant species because of the rapid sowing of superior modern varieties with high production, the clearing of new land, and the development of settlements. Gene banks are expected to ensure the sustainability of plant genetic resources.

For germplasm to be more efficient, it is necessary to implement more dynamic conservation strategies, such as in situ and on-farm conservation. Wild plant species should be preserved in national parks, and local plants should be preserved by communities, including farmers and non-governmental organizations, in certain areas for further development. On-farm conservation of traditional (local) varieties is a type of in-situ conservation. This conservation is dynamic because, besides being preserved, farmers can also cultivate these varieties. The morphology and appearance of plants can be changed by natural factors such as mutation and selection; however, this is not a concern, and farmers will carry out selection and management. Thus, on-farm conservation must be regarded as a component of a global strategy to preserve the germplasm that can be implemented either statically or dynamically.

The Role of Climate in Increasing Agricultural Productivity

Both abiotic and biotic factors determine the success of plant cultivation. Biotic factors are associated with the presence or absence of pests and plant diseases, whereas abiotic factors are associated with the plant-growing environment, specifically soil and climate. Climate plays a major role in plant life. Climate elements are determined by climate control factors, such as latitude, height above sea level, position relative to the ocean, low- and high-pressure centers, air mass movement, and obstructions by mountains and ocean currents. There are three types of climates on Earth: macro, meso, and microclimate. The microclimate is the most significant environment for plant life, which determines the success of plant cultivation in terms of harvest quality and quantity.

Climate plays an important role in supporting plant growth and production. Significant changes in climatic components, such as shorter rainfall distributions with high intensity and inconsistent distribution, can negatively impact crop yields (Hadi and Chalil, 2018). Several studies have shown that climate change has adverse effects on agricultural production. Among the climatic variables, rainfall, temperature, and humidity greatly influence plant growth.

Climate change can reduce agricultural production by 5–20%, threatening food security and the agricultural sector as a source of sustenance in Indonesia, which is classified as an agricultural country. As a result, the contribution of the agricultural sector to Indonesia's economic growth is decreasing due to the impact of climate change, which is caused, among other things, by increased greenhouse gas emissions that are a consequence of human activities that promote an increase in the Earth's temperature. An increase in the greenhouse gas content results in a greenhouse effect that inhibits the release of heat into the atmosphere, thereby increasing the temperature of the earth (Kweku et al. 2018; Anderson et al. 2016).

Global climate change, which occurs owing to changes in the pattern and intensity of climate elements over an extended period, has a negative impact on agricultural productivity. The average global temperature on ocean and land surfaces has increased by 0.85°C, increasing from 0.65°C to 1.06°C (IPCC, 2019). The increase in air temperature because of climate change affects atmospheric humidity and dynamics, as well as changes the rainfall pattern, resulting in an increase in the intensity and frequency of extreme climate anomaly events such as El -Niño and La -Niña (Fajri et al. 2019). El Niño prolongs the dry season, resulting in drought stress, whereas La Niña causes excessive rainfall, resulting in an increase in the likelihood of flooding. El Niño and La Niña cycles occurred every 5–7 years approximately 30–60 years ago, but they have been observed every 2–3 years for the past 5 years (Subagyono et al. 2014). Another impact of climate change is the increase in hydrometeorological disasters that are closely related to climate, such as floods, hurricanes, landslides, and droughts.

Climate change has continuous, discontinuous, or permanent effects. Increased air temperature, changes in rainfall patterns, increased groundwater salinity, decreased productivity, and changes in cropping patterns and indices are continuous impacts of climate change. Discontinuous impacts include crop failure, an increased frequency of extreme climate events such as floods, droughts, storms, typhoons, and outbreaks of pests and plant diseases. Meanwhile, the permanent impact is a reduction in agricultural areas owing to a decrease in sea level (Widiarta, 2016).

The seasonal shifts that result from the combined impact of continuous, discontinuous, or permanent effects of climate change influence agricultural productivity by affecting the planting patterns, planting time, planting index, harvest area, production, and quality of agricultural products. Additionally, the growth of plant pest organisms is exacerbated by the increasing temperature and humidity fluctuations caused by climate change, making it difficult for farmers to determine planting calendars and harvest periods. Climate change is a strategic issue that must be mitigated to minimize its adverse impact on agricultural productivity in Indonesia. Therefore, in the context of planning land use allocation, recommending planting patterns, expanding planting areas, and setting planting schedules/calendars, information on climate suitability is necessary. Each plant has a growth tolerance limit in terms of the climate and soil characteristics. Based on the growth conditions that are affected by the impacts of climate change, it is necessary to take strategies and initiatives in crop cultivation technology to increase crop productivity.

North Sumatra is located at 1–4° North Latitude and 98–100° East Longitude, with a unique geographical location flanked by the Strait of Malacca and the Indian Ocean and traversed by the Bukit Barisan mountains, which stretch from North to South. This geographical location greatly influences the dynamic weather patterns in North Sumatra. North Sumatra has a tropical climate influenced by passive winds and monsoons. Rainfall ranges from 800 to 4,000 mm/year, solar radiation is 43%, and the average air humidity is 78–91%. North Sumatra has agroclimatic zones based on Oldeman's classification, classified as types A, B1, C1, D1, D2, E1, and E2.

Cultivation Technology to Increase Germplasm Productivity in North Sumatra

Increasing the productivity of superior germplasm commodities in North Sumatra, such as rice, corn, cassava, sweet potatoes, soybeans, chilies, shallots, and garlic, requires appropriate cultivation technology to produce these crop commodities. The aim of developing cultivation technology is to create a conducive environment for plants, increase plant tolerance, and allow them to adapt to stresses, including climate change (Widiarta, 2016). Food and horticultural crops are classified as plants that are very vulnerable to the impacts of climate change, therefore, they must receive priority in preparing action plans for mitigating and anticipating climate change.

To anticipate climate change, it is essential to develop various technologies. These include plant types and varieties that are tolerant to abiotic stresses such as rising temperatures, drought, flooding, salinity, and resistant to biotic stresses like pests and diseases. Additionally, advancements in soil and plant management technologies, to enhance plant adaptability; and developing adaptive water management technology to address climate change, such as jug irrigation, drip irrigation, intermittent irrigation, and rotational dribble systems is also crucial to address climate change challenges (Widiarta, 2016).

Plant cultivation technology must adapt to plant needs and specific locations through the application of Good Agricultural Practices (GAP). Several factors that must be considered when increasing agricultural productivity in North Sumatra based on food security are:

1. Use of certified superior varieties

Superior varieties are an input in the plant production process, as strains resulting from plant breeding have special advantages, such as high yield potential, resistance to plant pests, tolerance to environmental stress, good product quality, and other characteristics, and have been released by the government.

The quality of agricultural production is improving because of the use of certified seeds. The use of certified seeds ensures better physical, genetic, and physiological quality. The significance of using quality and certified seeds is evident in their ability to enhance the value of products, reduce the risk of production failure and farming costs, and serve as a technical barrier to the invasion of imported seeds. Additionally, they provide timely, easy-to-track, and easy-to-obtain information on varieties that are more adaptive to certain growing environments.

2. Improve soil fertility

Technology to physically, chemically, and biologically improve soil fertility is required to provide input in the form of organic and biological fertilizers, as well as lime/dolomite to increase soil pH.

Soybean plants are legumes that can form symbiotic relationships with nitrogen fixing bacteria (rhizobia) in nodules in their roots, thereby increasing N availability for plants. Inoculation with *Rhizobium japonicum* biofertilizer (200 g/40 kg seeds) is required on land that has never been planted with soybeans.

Research has shown that the decrease in organic matter levels in rice fields is partly due to the continuous use of excessive amounts of organic fertilizers, which cause soil degradation. Additionally, the practice of transporting and burning straw also triggers a decline in rice field fertility. Burning straw and plant residues in rice fields plays a significant role in reducing air quality and soil nutrient content, which has an adverse impact on soil fertility. Rhofita (2016) reported that burning straw reduced N levels by 80%, P levels by 25%, K levels by 21%, and S levels 40-60%. Mandal et al (2004) reported that each planting season in rice cultivation will produce 7–10 tons of straw. Straw contains lignocellulose residues, including cellulose (30–45%), hemicellulose (20–25%), and lignin (15–20%). Straw also contains 44.71% organic C, 1.08% total N, 0.17% P, and 2.7% K (Nengah et al. 2023). According to Nengah (2020), the addition of straw compost can reduce the need for inorganic fertilizer by 20–80% and increase production equivalent to 100% inorganic fertilizer, thereby reducing the need for inorganic fertilizer and increasing the production of dry grain.

3. Water management

Water is a vital requirement for metabolic processes in plants. Plants require different amounts of water depending on the plant type and the growth and development phases of the cultivated plant. Cell growth is highly sensitive to water deficiency because of its effects on cell turgor, protein synthesis, and cell walls. In general, plants undergo three critical phases of water requirement: germination, flowering, and seed/pod filling. If water is unavailable during these critical phases, plant production decreases.

Water management aims at efficient and cost-effective water use and crop production, equitable distribution of water, and sustainability of the water resource use system. In rice plants, water management, which includes technical irrigation, limited irrigation, and rain-fed rice fields, needs attention. Different modes of irrigation in rice fields can influence planting patterns in rice fields. Lowland rice is a plant that is tolerant to waterlogging conditions due to the presence of aerenkim tissue, whereas soybeans are very sensitive to waterlogging conditions, although there is now a superior soybean variety that is tolerant to water saturation (Deja variety). Corn, cassava, sweet potatoes, chilies, shallots, and garlic are highly sensitive to flooding.

4. Fertilization

Fertilization is the process of providing nutrients to the soil, either directly or indirectly, to improve soil conditions, increase soil fertility, provide nutrients to plants, and improve plant production and quality. Fertilizer can be applied through the roots by spreading it, dropping it onto the plants, or spreading it between rows of plants. Fertilization can also be performed on leaves by spraying.

Fertilization must be carried out in a balanced manner, with the aim of increasing production, product quality, and fertilization efficiency; maintaining soil fertility; and avoiding environmental pollution and toxicity to plants. The concept of balanced fertilization involves providing a specific amount of fertilizer so that balanced essential nutrients are available and optimal in the soil. This is done in accordance with the soil's nutrient status, ensuring that plant needs, and targeted production (yield) can be achieved according to local conditions (specific location). Well-managed organic materials and biological fertilizers play an important role in increasing the effectiveness and efficiency of inorganic fertilizers. Fertilization must be performed in four precise ways by ensuring the right type, dose, time, and method of fertilization.

Balanced fertilization focuses on nutrient status in the form of a nutrient balance between the nutrients added and those taken up by the plant. Therefore, balanced fertilization and the prediction of nutrient requirements can be achieved by focusing on the P and K nutrient status maps, planting calendars for preparing fertilizer requirements, location-specific fertilizer recommendations, software, and soil testing equipment to prepare location-specific fertilizer recommendations (Kasno, 2019).

5. Control of plant pests and diseases

Plant pests and diseases have direct and indirect effects on crop yield. FAO reports that in 2021, pest attacks and plant diseases could reduce crop production by up to 40%. Integrated pest management (IPM) is the most effective method for pest control, as it ensures that all components could thrive without causing harm to cultivated plants. The aim of IPM is to minimize the use of chemical pesticides by integrating a variety of biological control techniques and chemical applications when other control techniques are unable to suppress pest populations. The IPM concept is based on the strategy of synergizing all compatible pest and disease control methods in accordance with ecological and economic principles. The operational principles of IPM include cultivating healthy plants, biological control, using resistant varieties, physical control, semi-chemical control by utilizing natural chemical compounds produced by certain organisms that influence the nature of insect pests, preserving and utilizing natural enemies, genetic control, and the use of pesticides if other controls are no longer effective in suppressing pest populations based on monitoring the control threshold for the target pest.

In soybean plants, IPM techniques include the use of resistant varieties, crop rotation, planting insect trap plants, simultaneous planting, biological control using natural enemies (parasitoids, predators, and pathogens), insect pests, vegetable pesticides, and lastly, chemical pesticides if the control methods used were no longer effective at suppressing pest populations. The results showed that biological control using the JTM 97C virus isolate was effective in controlling armyworms, leaf roller pests, and pod borers, with a mortality rate of 72–100%. The use of neem powder as a botanical pesticide containing azadirachtin, salanin, meliantriol, nimbin, and nimbidin is effective in controlling armyworms and whiteflies (Indiati and Marwoto, 2017; Acda, 2014; Bejo, 2012).

The control of plant diseases must also adhere to the principle of environmental balance, since it is inseparable from human activities that manipulate the environment, which in turn influences the development of disease. Environmental components are sought to have a harmonious and integrated influence on suppressing disease development. In nature, the development of disease occurs through an interaction between the pathogen that causes the disease, the host plant, and the environment (the disease triangle). Humans play an important role in manipulating these three components in the cultivation system; therefore, disease control based on the epidemic concept must be in accordance with the conditions and problems at each location and should prioritize the cultivation engineering process and environmental manipulation. In rice plants, this can be accomplished by planting resistant varieties, planting simultaneously and monitoring the application of technology in the field, which can also reduce the costs of pesticide application and other measures for controlling pest and plant disease. The results showed that there was a 20–50% increase in the production of rice planted simultaneously, especially during pest and plant diseases attacks. In rice fields of 600–1,000 ha, simultaneous planting can reduce the cost of pesticide use by up to 30% (Baehaki, 2009; Nuryanto, 2018).

Harvesting and Post-Harvest

Harvesting and postharvest handling are strategic activities that play an important role in effective product management and support food security. Post-harvest handling plays a role in minimizing yield losses, maintaining product quality, and enhancing both the added value and the income of farmers (Samaddar et al. 2017; Falatehan et al. 2021). Loss of harvest yield is influenced by various factors, including rice plant variety, harvest age, location, climatic conditions and seasons, harvesting tools, and post-harvest handling (Kumar and Kalita, 2017). The level of crop and postharvest loss was >10% for rice, >5% for corn, and >15% for soybeans. Therefore, strategies to reduce yield loss (weight loss), increase yield quality and competitiveness (standardize processes and results), and promote farmer welfare are necessary strategies to improve and handle post-harvest conditions. The participation of all individuals involved in harvest and post-harvest activities is needed to implement standardized harvest processes and results according to their respective functions (Molenaar, 2020).

Yield losses during post-harvest corn can be qualitative (decreased yield quality due to damaged grains, germinated grains, or wrinkled seeds during the drying, shelling, transportation, and storage processes) or quantitative (yield losses due to grains lost or left behind in the field during harvest).

In rice plants, post-harvest activities include harvesting, threshing, drying, storing, and milling the grain into rice. The primary challenges in the postharvest rice handling include high yield losses, low-quality grains produced by farmers, and inappropriate harvest times. Harvesting rice before it reaches physiological maturity compromises the quality of the produced rice due to the high percentage of green and calcified grains. Limited harvesting, threshing, and drying tools also result in longer harvest process chains and postharvest handling. Delays in harvesting often lead to delays in threshing and drying, owing to limited equipment for threshing and drying rice (Nugraha, 2012; Swastika, 2012; Molenaar, 2020).

Determining the appropriate harvest time is critical in the initial stage of postharvest handling. Rice plants ripen in four stages: milk, yellow, full, and dead. The appropriate harvest criteria for rice plants are as follows: 95% of the grain turned yellow, the flag leaves dried, the optimal age of the panicles was 30–35 days after flowering, the water content was 21–26%, and grain loss was approximately 16–30% (Kobarsih and Siswanto, 2015; Yahman et al. 2020; Sumarni and Syahbuddin, 2016).

Conclusion

North Sumatra has local plants that are well adapted to its environment and have been cultivated for generations. These local food and horticultural plants have distinctive morphological and quality characteristics. Germplasm resources play an important role in supporting food sovereignty, nutritional adequacy, and global market competitiveness. Protecting agricultural genetic resources and achieving the Sustainable Development Goals for food system resilience is increasingly important. Preserving plant germplasm, especially the diversity of endemic plants in North Sumatra, is essential. This can be done through *ex situ*, *in situ*, and on-farm conservation methods.

Increasing the productivity of superior germplasm commodities in North Sumatra such as rice, corn, cassava, sweet potatoes, soybeans, chilies, shallots and garlic needs to refer to cultivation technology in producing these crop commodities. Food and horticultural crops are classified as plants that are very vulnerable to the impacts of climate change, therefore they must receive priority in preparing action plans for mitigating and anticipating climate change.

To anticipate climate change, it is necessary to develop several technologies, such as developing plant types and varieties that are tolerant of abiotic environmental and biotic stresses, development of soil and plant management technology to increase plant adaptability and development of adaptive water management technology to climate. The development of plant cultivation technology must be adapted to plant needs and specific locations through the application of Good Agricultural Practices (GAP).

Disclaimer

None.

References

Acda, M. N. 2014. Repellent effects of Annona crude seed extract on the Asian subterranean termite *Coptotermes gestroi* Wasmann (Isoptera: Rhinotermitidae). *Sociobiology*, 61(3): 332–337.

Anderson, T. R., Hawkins, E., Jones, P. D. 2016. CO2, the greenhouse effect and global warming: from the pioneering work of Arrhenius and Callendar to today's Earth System Models. *Endeavour*, 40(3): 178–187.

Baehaki, S. E. 2009. Strategi pengendalian hama terpadu tanaman padi dalam perspektif praktek pertanian yang baik (Good Agricultural Practices). *Pengembangan Inovasi Pertanian*, 2(1): 65-78.

Bejo. 2012. Peningkatan efektivitas Helicoverpa armigera Nuclear Polyhedrosis Virus dengan beberapa bahan pembawa untuk mengendalikan hama polong kedelai. *Buletin Palawija*, 23: 38–43.

Chaniago. N. 2017. Karakteristik Morfologi Beberapa Kultivar Padi Gogo Lokal Sumatra Utara. *Agrica Ekstensia*, 11(2): 46-54.

Currah. 2022. *Allium. Crop Science: Recent Advances p. 101–118* CABI Publishing. USA.

Darwis, V. 2018. Potensi kehilangan hasil panen dan pasca panen jagung di Kabupaten Lampung Selatan. *Journal of food system and agribusiness* 2(1): 55–67. http://dx.doi.org/10.25181/jofsa.v2i1.1054.

Ekawati, I. 2010. Pengembangan Potensi Jagung Menuju Revitalisasi Pertanian di Kabupaten Sumenep. *Jurnal Pertanian CEMARA*, 7(1): 25–33.

Fajri, H. C., Siregar, H., Sahara, S. 2019. Impact of climate change on food price in the affected provinces of el nino and la nina phenomenon: Case of Indonesia. *International Journal of Food and Agricultural Economics* (IJFAEC), 7(4): 329–339. http://dx.doi.org/10.22004/ag.econ.296761

Falatehan, A. F., Syaukat, Y., Hastuti, H., Nasrullah, N. 2021. Paddy Loss and Its Implication to Fertilizer Subsidy in Indonesia. HAYATI *Journal of Biosciences* 28(1): 73–73.

FAO, IFAD, UNICEF, WFP and WHO. 2018. *The state of food insecurity in the world 2018: Building climate resilience for food security and nutrition.* Rome, FAO. http://www.fao.org/docrep/018/i3434e/i3434e.pdf. Accessed on: 12th April 2024.

FAO. 1997. *The State of The World's Plant Genetic Rsources for Food and Agriculture. Food and Agriculture of The United Nations.* Rome (Italy): FAO. pp. 540.

FAO. 2014. *Genebank Standards for Plant Genetic Resources for Food and Agriculture. Rev.* ed. Rome (Italy): FAO. pp. 182.

FAO. 2016. *National Level Conservation and Use of Farmers' Varieties/Landraces: Revised Draft Voluntary Guidelines*. CGRFA-16/17/Inf.18.

Hadi, H., Chalil, D. 2018. Effect of Irrigation on the Changes in Wetland Rice Productivity (*Oryza sativa* L.) due to a Climate Change in North Sumatra Province. *Indonesian Journal of Agricultural Research*, 1(1): 51–57. https://doi.org/10.32734/injar.v1i1.183.

HLPE. 2017. Nutrition and food systems. A report by the high-level panel of experts on food security and nutrition of the committee on World Food Security, Rome.

Indiati, S. W., Marwoto. 2017. Penerapan Pengendalian Hama Terpadu (PHT) pada Tanaman Kedelai. *Buletin Palawija,* 15 (2): 87–100, doi:10.21082/bul palawija. v15n2.2017.p87-100.

IPCC. 2019. IPCC Special Report on Climate Change, Desertification, Land Degradation, Sustainable Land Management, Food Security, and Greenhouse gas fluxes in Terrestrial Ecosystems. *Intergovernmental Panel on Climate Change*.

Kasno, A. 2019. Perbaikan tanah untuk meningkatkan efektivitas dan efisiensi pemupukan berimbang dan produktivitas lahan kering masam. *Jurnal Sumberdaya Lahan* 13(1): 27–40.

Kobarsih, M., Siswanto, N. 2015. Penanganan susut panen dan pasca panen padi kaitannya dengan anomali iklim di wilayah Daerah Istimewa Yogyakarta. *Planta Tropika*, 3(2): 100–106.

Kumar, D. Kalita, P. 2017. Reducing postharvest losses during storage of grain Crops to strengthen food security in developing countries. *Foods*, 6:1–22.

Kweku, D. W., Bismark, O., Maxwell, A., Desmond, K. A., Danso, K. B., Oti-Mensah, E. A., Asenso, T., Adormaa, B. B. 2018. Greenhouse effect: greenhouse gases and their impact on global warming. *Journal of Scientific research and reports*, 17(6): 1–9. https://doi.org/10.9734/JSRR/2017/39630.

Mandal, K. G., Misra, A. K., Hati, K. M., Bandyopadhyay, Mohanty, P. M. 2004. Rice residue-management options and effects on soil properties and crop productivity. *Food, Agriculture & Environment*, 2 (1): 224–231.

Molenaar, R. 2020. Panen dan pascapanen padi, jadung dan kedelai. *Eugenia*, *26*(1):17–28.

Nengah, M. I. 2020. Pemanfaatan Kompos Jerami Padi Guna Memperbaiki Kesuburan Tanah dan Hasil Padi. *Rona Teknik Pertanian*: 13:2;2020.

Nengah, M. I., Sukmadewi, D. K. T. S. 2023. Improving Soil Fertility Through Agricultural Waste Treatment in Subak Telun Ayah, Gianyar, Bali. *Asian Journal of Community Services*, *2*(3): 235-246.

North Sumatra Province. 2021. Rencana Kerja Pemerintah Daerah (RKPD) Provinsi Sumatra Utara Tahun 2021.

Nugraha, S. 2012. Inovasi Teknologi Pascapanen untuk Mengurangi Susut Hasil dan Mempertahankan Mutu Gabah/Beras di Tingkat Petani. *Buletin Teknologi Pascananen Pertanian*, Vol 8 (1): 48–61.

Nuryanto, B. 2018. Pengendalian penyakit tanaman padi berwawasan lingkungan melalui pengelolaan komponen epidemik. *Jurnal Penelitian dan Pengembangan Pertanian*, *37*(1): 1-12.

Reardon, T., Timmer, P. 2012. The economics of the food system revolution. *Annual Review of Resource Economics*, *4*, 225–264.

Samaddar, A., Azam, M. M., Singaravadivel, K., Venkatachalapathy, N., Swain, B. B., Mishra, P. 2017. Postharvest management and value addition of rice and its by-products. In *The future rice strategy for India* pp. 301–334. Academic Press.

Statistics of Sumatera Utara, 2024. Number of Population by Type of Sex and Regency/City (People), 2021-2023. https://sumut.bps.go.id/indicator/12/65/1/jumlah-penduduk-menurut-jenis-kelamin-dan-kabupaten-kota.html. Accessed on : 12th April 2024.

Statistics of Sumatera Utara. 2021. *Vegetable Crop Production by Regency/City and Type of Crop in North Sumatra Province* (Quintals) 2021. https://sumut.Badan Pusat Statistik.go.id/indicator/55/541/1/produksi-tanaman-sayuran-menurut-kabupaten-kota-dan-jenis-tanaman-di-provinsi-Sumatra-utara-kwintal-.html.

Statistics of Sumatera Utara. 2022[a]. *Harvest Area, Production, and Average Yield of Corn by Regency/City 2020–2022.* https://sumut.BadanPusatStatistik.go.id/indicator/53/156/1/luas-panen-produksi-dan-rata-rata-produksi-jagung-menurut-kabupaten-kota.html.

Statistics of Sumatera Utara. 2022[b]. *Harvest Area, Production, and Average Yield of Sweet Potatoes by Regency/City 2020–2022.* https://sumut.BadanPusatStatistik.go.id/indicator/53/203/1/luas-panen-produksi-dan-rata-rata-produksi-ubi-jalar-menurut-kabupaten-kota.html.

Statistics of Sumatera Utara. 2022[c]. *Harvest Area, Production, and Average Yield of Cassava by Regency/City 2020–2022.* .https://sumut.Badan Pusat Statistik.go.id/indicator/53/201/1/luas-panen-produksi-dan-rata-rata-produksi-ubi-kayu-menurut-kabupaten-kota-.html.

Subagyono, Kasdi, Surmaini, E. 2014. "Pengelolaan Sumberdaya Iklim Dan Air Untuk Antisipasi Perubahan Iklim." *Jurnal Meteorologi dan Geofisika*, 8(1).

Tendall, D. M., Joerin, J., Kopainsky, B., Edwards, P., Shreck, A., Le, Q. B., Kruetli, P., Grant, M., Six, J. 2015. Food system resilience: Defining the concept. *Global Food Security*, 6:17–23.

Widiarta, I. N. 2016. Teknologi pengelolaan tanaman pangan dalam beradaptasi terhadap perubahan iklim pada lahan sawah. Jurnal Sumberdaya Lahan, 10(2): 91–102.

Yahman, M. B., Widada, D., Profita, A. 2020. Analisis risiko dan penentuan strategi mitigasi pada proses produksi beras. MATRIK: *Jurnal Manajemen dan Teknik Industri Produksi*, 20(2): 67–78.

Chapter 15

The Current Status and Research Directions of Renewable Energy in North Sumatra

Irvan[1,2,*]
Seri Maulina[1,2,3]
and Renita Manurung[1,2]
[1]Chemical Engineering Department, Faculty of Engineering,
Universitas Sumatera Utara, Medan, Indonesia
[2]Sustainable Energy and Biomaterial Center of Excellence,
Universitas Sumatera Utara, Medan, Indonesia
[3]Research Cluster Life Cycle Assessment, Universitas Sumatera Utara, Medan, Indonesia

Abstract

As part of the national energy policy, renewable energy use has been implemented in Indonesia and has continuously improved to strengthen the national energy security and reduce national emissions. To that end, the Indonesian government has created a General National Energy Plan (GNEP, in Bahasa we call RUEN) with a new renewable energy mix target of 23% by 2025. In connection with this, the North Sumatra Provincial Government has prepared a General Regional Energy Plan, (GREP, RUED in Bahasa) with a renewable energy mix target of 33% by 2025. The potential renewable energy resources in North Sumatra include hydroelectric, biomass, solar and geothermal energy. Hydroelectric energy is generated through large-, mini-, and micro-scale power plants. Meanwhile, biomass energy is derived from rice husks, coconut shells, palm shells and sugar cane bagasse. North Sumatra has potential resources to produce crude palm oil (CPO) by

* Corresponding Author's Email: irvan@usu.ac.id.

In: Natural Resources, Environment, and Food and Energy Security …
Editors: Mohammad Basyuni and Severino Salmo III
ISBN: 979-8-89530-283-5

processing up to 11 million tons of fresh palm fruit bunches (FFB), which generate waste in the form of empty bunches, fibers, shells, and liquid waste (palm oil mill effluent, POME). Palm oil solid waste, such as fiber, empty palm fruit bunches, and shells, is used as a boiler fuel in several palm oil mills (POM) in North Sumatra. In addition to solid waste, several mills have used liquid waste for fermentation into biogas, which is then converted into electricity. The Ministry of Energy and Mineral Resources (MEMR, ESDMin Bahasa) has issued regulations regarding the use of renewable energy sources to provide electricity. The State Electricity Company (PT PLN in Bahasa) as the buyer and private developer as the seller can negotiate the price of electricity from all renewable energy sources. The goals of the national energy policy can be realized upon the agreement of all parties supporting it.

Keywords: renewable energy, North Sumatra, palm oil mill effluent, biodiesel, biogas, electricity

Introduction

North Sumatra has vast and beautiful natural landscapes and enormous renewable energy potential. The potential renewable energy resources in North Sumatra include hydroelectric, solar, and geothermal. According to the New Renewable Energy Information and Investment Services and Energy Conservation (Lintas EBTKE in Bahasa) website of the Ministry of Energy and Mineral Resources, the renewable energy potential in North Sumatra is approximately 19 GW, comprising 3,808 MW for water energy, 1,204 MW for mini- and micro-hydro electric energy, 11.85 GW for solar energy, 356 MW for wind energy, and 2,206 MW for geothermal energy.

Geothermal Energy

North Sumatra has impressive geothermal potential, which is suitable for increasing the use of renewable energy. Table 1 presents a list of the geothermal working areas and geothermal potentials in North Sumatra. Table 2 lists the current progress of geothermal energy in North Sumatra, with three operating geothermal power plants (PLTPs) with a total installed capacity of 561.4 MW.

Table 1. Geothermal Potential in North Sumatra (Kadri et al. 2023)

Location	Potential (MW)
Sorik Marapi-Roburan-Sampuraga	200
Sipoholon Ria Ria	172
Simbolon Samosir	150
Gunung Sibual-Buali	1,146
Gunung Sibayak–Gunung Sinabung	114
Dolok Marawa	88
Pincurak	50
Sibubuhan	100

Table 2. Operating Geothermal Power Plants (PLTPs) in North Sumatra (Tefbana and Situmorang, 2024)

Name	Installed Capacity (MW)	Company
PLTP Sibayak	12	PT Pertamina Geothermal Energy
PLTP Sarulla	330	PT Pertamina Geothermal Energy and Joint Operation Contract Sarulla Operation Limited
PLTP Sorik Marapi	99,4	PT Sorik Marapi Geothermal Power

Considering its impressive geothermal potential, the province continues to play a key role in maintaining the energy balance in Indonesia. Moreover, the exploration of North Sumatra reflects hope for a sustainable future.

Hydroelectric Power

North Sumatra plays an important role in the hydroelectric power generation sector. Several ongoing and planned projects aim to exploit the potential of the region to meet local and national energy needs. In particular, large-scale dams, river projects, and micro-hydroelectric power systems are under construction, contributing to the energy diversification and economic growth of North Sumatra. Such hydroelectric power plants have a wide spectrum of electricity capacities, from 50 MW to 317 MW. Table 3 lists the best five operating hydroelectric power systems in North Sumatra, with a total installed capacity of 915 MW.

Indonesia has taken measures to diversify its energy mix and reduce its dependence on fossil fuels. Hydroelectric power offers an excellent alternative for reducing the country's dependence on imported fuel and increasing its energy security. By exploiting its enormous hydropower potential, North Sumatra could contribute significantly to the national energy network and strengthen Indonesia's energy independence.

Table 3. Operating Hydroelectric Power Plant (PLTAs) in North Sumatra (Wahyuni, 2024)

Name	Installed Capacity (MW)
PLTA Tangga	317
PLTA Sigura-gura (Asahan II)	286
PLTA Sigura-gura (Asahan I)	180
PLTA Lau Renun	82
PLTA Sipansihaporas	50

Biogas

As the one of the provinces with the largest number of palm oil plantations in Indonesia, North Sumatra has enormous potential for biogas development. Fresh palm fruit bunches are used as a renewable energy source; however, approximately 60% of their quantity is wasted. Palm oil waste has the bioenergy potential of 57,000 MW. Based on cooperation schemes with other parties, such as the build operate transfer (BOT) scheme, several palm oil mills have built biogas plants (PLTBgs) using from palm oil mill effluents. Table 4 lists various operating PLTBgs in North Sumatra.

Table 4. Biogas Plant (PLTBgs) in North Sumatra (Hartatik, 2022)

Name	Capacity	Company
PLTBg Kwala Sawit	1 MWe	PTPN II and PT Pertamina Power Indonesia
PLTBg Pagar Merbau	1 MWe	PTPN II and PT Pertamina Power Indonesia
PLTBg Pasir Mandoge	2 MWe	PTPN IV and PT Karya Mas Energi (KME)
PLTBg Sei Mangkei	2.4 MWe	PTPN III and Pertamina NRE
PLTBG Hapesong	0.8 MWe	PTPN III
PLTBG Ukindo	0.8 MWe	PT United Kingdom Indonesia Plantations
PLTBg Tasik Raja	2 MWe	PT. Tasik Raja
PLTBg Siringo-ringo	1.5 MWe	PT. Siringo-ringo
PLTBg Gunung Melayu-1	1.2 MWe	PT. Saudara Sejati Luhur
PLTBg Negeri Lama-2	1.2 MWe	PT. Hari Sawit Jaya
PLTBg Supra Matra Abadi	1.2 MWe	PT. Supra Matra Abadi
PLTBg Indo Sepadan Jaya	1.2 MWe	PT. Indo Sepadan Jaya
PLTBg Gunung Melayu	1.2 MWe	PT. Gunung Melayu
BioCNG Langkat	300 MMBTU biogas	PT United Kingdom Indonesia Plantations

Table 4 shows that the largest use of biogas is achieved by converting it into electricity. This is logical because the biogas produced has low methane purity and low pressure. With a low and uncompressed methane gas concentration, this type of biogas can only be used to generate electricity to fulfill its own needs. Such biogas is not efficiently transported to other locations owing to its large volume. Compressed biogas is also an important component with the national production target of 490 million m^3 by 2025. Currently, national biogas production has reached around 100 million m^3.

Biomass

In 2019, the Riau, East Java and North Sumatra provinces were ranked the highest in terms of biomass energy production potential. In particular, North Sumatra accounts for 1,927 MW of the 2,895 MW biomass energy comes from the palm oil industry. Table 5 lists the different types of potential biomass in North Sumatra.

The development of power plants using biomass fuel as a renewable energy source was first recorded on a commercial scale in 2010. The first development began in the field of industrial fields from the Medan Industrial Region (KIM) 1 and KIM 2, followed by KIM 3 in 2013. Table 6 presents a list of the biomass developments in North Sumatra from 2010 to 2021.

Table 5. Types of Potential Biomass in North Sumatra (Irwanto et al, 2021)

Material	Electrical Potential (MW)
Palm oil	1,927
Sugar cane	30
Rubber wood	220
Coconut	5
Rice husk	472
Corn	111
Cassava	11
Wood	4
Livestock	16
Garbage	99
Total	2,895

Various technologies have been used in the power generation of biomass-fueled electricity, such as direct combustion of coal, conversion of refuse-derived fuel, gasification, sanitary landfills, and incinerators.

Currently, municipal waste management using cofiring technologies constitute the largest portion of biomass-based power generation.

Table 6. List of Biomass Development in North Sumatra from 2010 to 2021 (Ardiansyah and Ekadewi, 2022)

Name	Year	Resources	Capacity (MW)	Company
PLTBm North Sumatra, KIM I	2010	Palm shells, sugar cane, rice husks, corncobs, tapiocadregs, and sawdust	30	PT. Growth Sumatra Industry
PLTBm North Sumatra, KIM II	2012	Palm shells, sugar cane, rice husks, corncobs, tapiocadregs, and sawdust	25	PT. Growth Sumatra
PLTBm Tangkahan, KIM III	2013	Palm shells, sugar cane, rice husks, corncobs, tapiocadregs, and sawdust	30	PT. Growth Asia
PLTBm Simalungun	2013	Palm shells, sugar cane, rice husks, corncobs, tapiocadregs, and sawdust	30	PT. Harkat Sejahtera
PLTBm North Sumatra, KIM I	2015	Palm waste	3	PT. Victorindo
PLTBm North Sumatra, KIM I	2015	Palm waste	1.8	PTPN III
PLTBm Tanjung Selamat	2020	Rubberwood	9,9	PT Cipta Multi Listrik Nasional
PLTBm Nias	2022	Bamboo	5	PT. Energi Infra Nusantara
PLTBm Victorindo Alam Lestari	2019	Palm waste	3.0 MWe	PT. Victorindo Alam Lestari
PLTBm Karya Agung Sawita	2019	Palm waste	1.0 MWe	PT. Karya Agung Sawita
PLTBm Rambutan	2019	Palm waste	1.8 MWe	PTPN III

Biodiesel

Indonesia produced 16.37 million kL of biodiesel in 2021. According to the data from the Central Statistics Agency, the largest biodiesel production was obtained from Riau Province, with a total volume of 5.1 million kL, followed by East Java at 3.88 million kL of biodiesel, East Kalimantan at 1.37 million kL, South Kalimantan at 1.01 million kL, and North Sumatra at 912.000 kL. On February 1, 2023, Indonesia implemented the use of diesel fuel mixed with 35% biodiesel from CPO or B35.

PT Perkebunan Nusantara III (Persero) or PTPN III is planning on building a biodiesel factory with a production capacity of 450 thousand tons per year in the Sei Mangkei Special Economic Zone (KEK), North Sumatra, with the production start target of early 2025. Table 7 lists the three best biodiesel producing plants in northen Sumatra.

Table 7. Biodiesel Plants in North Sumatra (Manurung et al. 2023)

Name	Production amount (kL)
PT Sintong Abadi	35,000
PT Musim Mas	459,770
PT Permata Hijau Palm Oleo	417,241

PT. Musim Mas and PT. Permata Hijau Palm Oleo are two Indonesian biodiesel industry companies that have been designated as suppliers of biodiesel to PT. Pertamina (Persero) and PT. AKR Corporindo Tbk. This designation was provided in the Decree of the Minister of Energy and Mineral Resources (Number 3756K/10/MEM/2017). The allocation of biodiesel from November 2017 to April 2018 to Pertamina by PT Musim Mas was 160,300 kL and that to PT Permata Hijau Palm Oleo was 50,482 kL. One of the subsidiaries of PTPN III (Persero), Palm Co., will start building a biodiesel production facility in the Sei Mangke Industrial Area, North Sumatra in 2024 to increase the number of companies producing biodiesel in North Sumatra (Manurung et al. 2024).

Currently, most raw biodiesel materials used in Indonesia come from CPO and other plants, including jatropha, candlenut, and *nyamplung* (*Calophyllum inopyllum*). Considering the great potential of North Sumatra as a CPO producer with a palm oil land farming area of approximately 1.3 million hectares, CPO-based biodiesel and its derivatives have the highest development potential; therefore, North Sumatra plays an important role in supporting raw biodiesel materials. In addition to CPO, used cooking oil is another raw material with high biodiesel potential.

Solar, Wind, and Nuclear Energy

Solar, wind and nuclear energy are sources of rapidly developing renewable energy sources worldwide. Unlike other countries, Indonesia is still lagging behind in the utilization of these energy sources. However, North Sumatra has tremendous potential for such renewable energy sources, as shown in Table 8.

Table 8. Potential of Solar, Wind, and Nuclear Energy in North Sumatra (Paundra and Nurdin, 2022)

Energy	Potential
Solar	Area: 72,981.23 km^2 Technical Potential: 11,851 MW
Wind	72,981.23 356 Area: 72,981.23 km^2 Technical Potential: 356 MW
Nuclear*	Uranium: 31,567 ton Thorium: 126,821 ton

*All Sumatra.

Environmental Impact in Renewable Energy

Environmental issues, such as global warming, ozone layer depletion, acidification, and eutrophication, which can cause damage to natural habitats and the extinction of biodiversity, have been debated in international forums. In particular, continuous evaluation of the ecological, economic, and social aspects of the materials, processing, and products must be conducted to achieve sustainable management. Several tools have been developed to support this strategy, including life cycle assessment (LCA), which is one of the most frequently used tools and has been standardized in ISO 14040, including the Principles and Framework, and ISO 14044 includes requirements and guidelines. LCA has been used in the sustainability framework by several business associations and companies in the industry. One of the steps in LCA is life cycle impact assessment (LCIA), which collects data on life cycle inventories and determines environmental impact values, by quantifying the energy and materials used and waste released into the environment.

In this article, the environmental impacts of the palm oil industry were examined as agro industries in North Sumatra. The data collected includes the energy consumption in nurseries, plantations and palm oil mills with the functional units of 1 kg main nursery seeds, 1 kg fresh fruit bunches, and 1 kg CPO. Data was collected monthly for a year. The energy utilized in the palm oil industry in North Sumatra includes electricity supplied by the state electricity company (PLN), which generally uses fossil fuels and diesel-powered generators. Table 9 presents the energy consumption in nurseries, plantations, palm oil mill-A and palm oil mill-B.

Table 9. Energy consumption in a Nursery, Plantation, Palm Oil Mill-A and Palm Oil Mill-B

Source of Energy	Unit	Nursery	Plantation	Palm Oil Mill-A	Palm oil Mill-B
Electricity	kWh	8,305.8	9,719.1	391,270.8	192,145.0
Fuel	L	4,128.1	2,875.1	6,235.8	13,311.9

According to the Environmental Products Declaration, the impacts of energy usage are categorized into acidification, eutrophication, global warming potential, photochemical oxidation, abiotic depletion, abiotic depletion (fossil fuels), water scarcity, and ozone layer depletion. Figure 1 shows the composition of these impact categories of energy usage in nurseries, plantations, and palm oil mills.

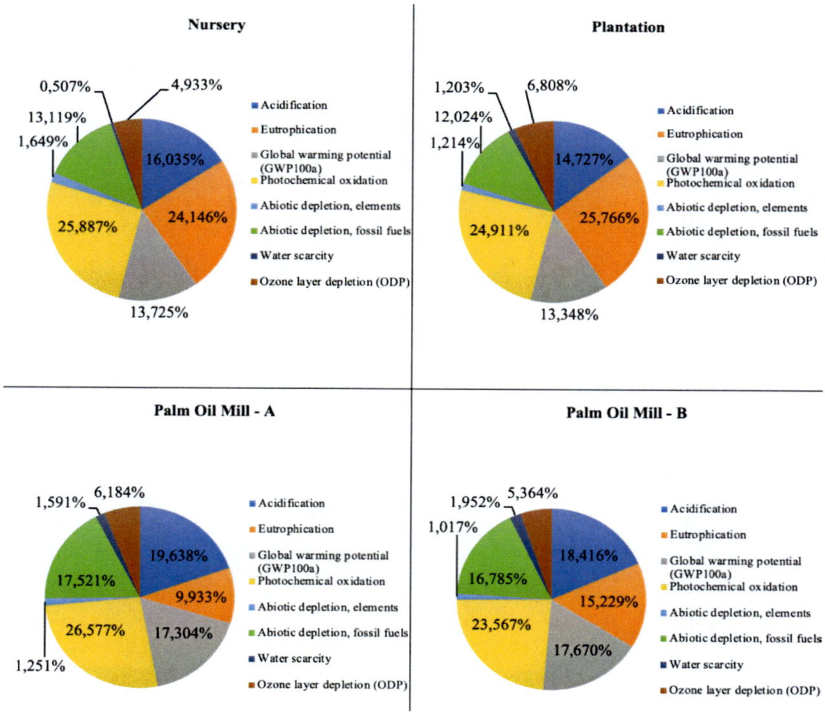

Figure 1. Impact of Energy Usage in a Nursery, Plantation, Palm Oil Mill-A, Palm Oil Mill-B.

As shown in Figure 1, photochemical oxidation, eutrophication, acidification, abiotic depletion (fossil fuels), and global warming potential constitutes a large portion of energy usage. Photochemical potential is a secondary air pollution caused mainly by the reaction of sunlight with emissions from fossil fuel combustion, creating other chemicals, such as ozone. Photo chemical potential has the most prominent impact of energy usage in nurseries, plantations, and palm oil mills A and B, because of the use of fossil fuels as energy sources. Eutrophication is a process where nutrients accumulate in a body of water, thereby increasing the growth of microorganisms that caused water blooms and bacterial growth that may deplete the oxygen in water. Nurseries and plantations have the highest impact on eutrophication because of their fertilizer usage. The acidification effect of burning fossil fuels is approximately equal for nurseries, plantations, and palm oil mills. The abiotic depletion of fossil fuels pertains to the depletion of non-renewable fossil fuel resources, expressed in MJ of fossil fuel. The abiotic depletion owing to energy usage in nurseries, plantations and palm oil mills is approximately equal.

Discussion

The threat of energy scarcity in the future and the environmental impact caused by fossil fuels have led researchers in North Sumatra to continuously innovate and develop new and renewable energy. Several study centers have been established under the command of Universitas Sumatera Utara, such as the Sustainable Energy and Biomaterial Center of Excellence and the Center Study of Volcanology. A Center for the Study of Volcanology was conducted considering the presence of volcanoes and hot springs, such as Mount Sibayak, Mount Sinabung, and Toba Caldera, in North Sumatra which needs to be investigated for their utilization.

For biogas research, several topics can be explored, such as upgrading biogas by removing impurities (CO_2 and H_2S), biogas liquefaction and transportation biogas, conversion of biogas to electricity or H_2, and hydrogen storage.

For the biodiesel field, commercial biodiesel production uses a transesterification reaction catalyzed by a homogeneous alkali catalyst, namely, NaOH or KOH, dissolved in methanol. However, homogeneous base catalysts have several disadvantages such as limited recovery, need for neutralization, unreusability, and the separation of fatty acid methyl esters (FAME) from the catalyst, and generation of liquid waste. Therefore, heterogeneous base catalysts have been developed for biodiesel production to overcome the limitations of homogeneous base catalysts. Heterogeneous base catalysts have various advantages, including their ease of separation, reusability, and environmental friendliness.

Heterogeneous catalysts can be obtained by impregnating metal catalysts onto a silica (SiO2) matrix using a wet impregnation method to obtain catalysts for biodiesel production. The impregnation of alkali metals on silica can increase the number of active sites on the catalyst, thereby increasing its conversion and yield. Palm leaves are one of the materials that can be used as a silica source in the preparation of sodium silica catalysts. In particular, palm leaf ash contains quite high levels of silica (70.11%) and metal oxides such as Al_2O_3, Fe_2O_3, CaO, MgO, and Na_2O. Therefore, palm leaf ash can be used as a silica source in the preparation of sodium silica catalysts.

Conclusion

1. The development of renewable energy in North Sumatra for each type is shown as follows:
   a. As of 2019, the total installed capacity for geothermal energy reached 561.4 MW, comprising 26.3% of the geothermal resources in Indonesia.
   b. As of 2019, the installed capacity of large-scale hydropower plants was 915 MW.
   c. In the biogas sector, a total capacity of 17.5 MWe plant was installed.
   d. In the biodiesel sector, 912,011 kL of biodiesel was produced in North Sumatra.
2. The solar, wind and nuclear energy sectors in North Sumatra have not been utilized for energy harvesting because of the large investments required. Moreover, the nuclear sector requires sophisticated technology, especially in terms of safety.
3. The North Sumatra Government has adopted various strategies to develop renewable energy, including requiring energy actors to utilize renewable energy, committing to implementing energy efficiency, and creating an energy saving culture.

Disclaimer

None.

References

Ardiansyah, H., Ekadewi, P. 2022. *Indonesia post-pandemic outlook: Strategy towards net-zero emissions by 2060 from the renewables and carbon-neutral energy perspectives* (133–158).

Hartatik. 2022. North Sumatra's 25 BioCNG Plants to be the Largest Energy Decarbonisation Project in Southeast Asia. *Energy Transition, Gas, Renewables.*

Irwanto, M., Purba, D. P., Nisja, I. 2021. Potential of Renewable Energy in North Sumatra Until 2028 Using LEAP. *Journal of Electronics, Telecommunication, Electrical and Physics Science* 1(1): 7-12. doi: https://doi.org/10.20122/etaps.v1i1.3077.

Kadri, M., Muztaza, N. M., Nordin, M. N. M., Zakaria, M. T., Rosli, F. N., Mohammed, M. A., Zulaika, S. 2023. Integrated geophysical methods used to explore geothermal potential areas in Siogung-Ogung, North Sumatra, Indonesia. *Bulletin of the Geological Society of Malaysia* 76(1): 47-53. doi: https://doi.org/10.7186/bgsm76202304.

Manurung, R., Hasibuan, R., Siregar, A. G. 2023. Preparation and characterization of lithium, sodium, and potassium silicate from palm leaf as a potential solid base catalyst in developed biodiesel production. *Case Studies in Chemical and Environmental* 9(2): 100543.

Manurung, R., Parinduri, S. Z. D. M., Hasibuan, R., Tarigan, B. H., Siregar, A. G. 2023. Synthesis of nano-CaO catalyst with SiO2 matrix based on palm shell ash as catalyst support for one cycle developed in the palm biodiesel process. *Case Studies in Chemical and Environmental* 7(3): 100345. doi: 10.1016/j.cscee.2023.100345.

Paundra, F., Nurdin, A. 2022. Study of the Potential and Development of Renewable Energy Power in Indonesia: A review. *Journal of science, Technology, Education and Mechanical Engineering* 3(2): 62-72.

Tefbana, F. A., Situmorang, B. 2024. Community in speed space: Media Information about Ulumbu PLTP geothermal. *Journal of Sumatera Sociological Indicators* 3(1): 253-259.

Wahyudi, N. A. 2024. New renewable energy: North Sumatra accelerates hydropower plant project. *Bisnis Indonesia - Energi baru terbarukan: Sumut akselerasi proyek PLTA.*

Chapter 16

Community Empowerment in Improving the Quality of Regional Potential for the Tourism Sector in North Sumatra

Tulus[1,*]
Rosdanelli Hasibuan[2]
and Elvina Herawati[1]
[1]Department of Mathematics, Universitas Sumatera Utara, Medan, Indonesia
[2]Department of Chemical Engineering, Universitas Sumatera Utara, Medan, Indonesia

Abstract

Community empowerment can be defined as a process concentrated on local communities, and it is an integral part of developing the tourism industry to ensure that local community members benefit. It is based on democratic participation and the critical reflection of members, allowing them to play an active role in decision-making regarding conditions and resources. North Sumatra, an Indonesian province, currently has three National Strategic Projects: The Lake Toba Tourism Area, the Sei Mangkei Special Economic Zone, and the Kuala Tanjung Industrial Area. North Sumatra has several destinations that are often visited by both national and foreign tourists. Well-known tourist destinations in North Sumatra include Lake Toba, Bukit Lawang, Tangkahan, Nias Island, Medan, and Berastagi. Problems in managing tourist destination areas usually arise from sustainability, namely, preserving nature and local culture. This article describes the community empowerment of the tourism sector in North Sumatra to

* Corresponding Author's Email: tulus@usu.ac.id.

In: Natural Resources, Environment, and Food and Energy Security …
Editors: Mohammad Basyuni and Severino Salmo III
ISBN: 979-8-89530-283-5

improve the quality of tourist area management and the benefits obtained by the community. The government needs to effectively form and implement laws, policies, and regulations and collaborate with other institutions. In the case of Lake Toba, laws, policies, and regulations must create a friendly environment to strengthen sustainable tourism in North Sumatra. Scholars' recommendations have also been found to significantly strengthen sustainable tourism in Lake Toba. In implementing the tri-dharma of higher education, community service programs are often conducted by lecturers in the community empowerment and tourism sectors.

Keywords: community empowerment, sustainable tourism, community service

Introduction

According to Law Number 8 of 2023, which pertains to North Sumatra Province, North Sumatra is an Indonesian province situated in the northern region of the Sumatra Island. Medan is the capital and largest city in the region. The region is adjacent to Aceh in the northwest and Riau and West Sumatra in the southeast. It has two distinct coastlines along the Indian Ocean and the Strait of Malacca and has a maritime boundary with Malaysia to the east. Figure 1 shows the area included in North Sumatra. North Sumatra is the fourth most densely populated province in Indonesia, followed by West, East, and Central Java. It is also the third largest province in terms of land area on the island of Sumatra, after the South Sumatra and Riau provinces. The area of 72,460.74 km2 is equivalent to the size of Sierra Leone, Scotland, or Maine.

North Sumatra has numerous international, national, and local tourist attractions. It has three ongoing National Strategic Projects, including the Lake Toba Tourism Area, the Sei Mangkei Special Economic Zone, and the Kuala Tanjung Industrial Area.

The tourism sector is one of the leading potential areas in North Sumatra. Lake Toba, the largest volcanic lake in the world, is a natural resource with undeniable tourism potential (Figure 2). North Sumatra's natural beauty holds great potential and has contributed significantly to the economy through tourism. Lake Toba's main attraction has made it a national priority tourist destination. To support becoming an international standard tourist destination, the central government continues to develop infrastructure, particularly to advance connectivity.

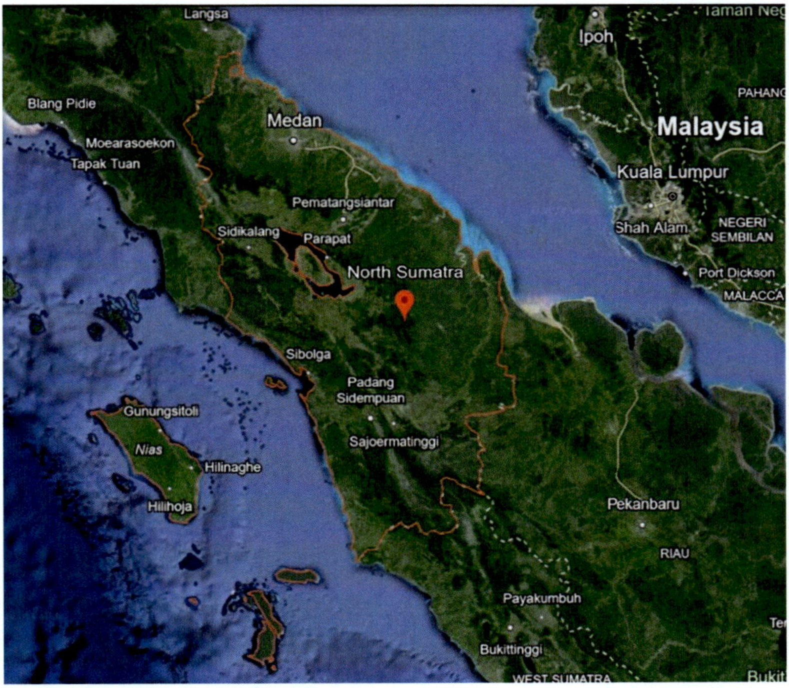

Source: https://earth.google.com.

Figure 1. The map of North Sumatra Province.

Figure 2. View of Lake Toba and Orang Utan in Bukit Lawang.

The tourism industry has the potential to generate revenue for national economies. Ginanjar and Runingsawitri (2023) propose the concept of Community Based Tourism (CBT) as a framework for managing tourism that focuses on engaging local communities. The implementation of CBT grants the community a central role in diverse tourism activities and ensures that it receives the greatest benefits. Rachmawati et al. (2021) analyzed the level of community engagement in tourism development. Purworini et al. (2023) discovered that a combination of community empowerment, tourism development, and promotional and marketing communication techniques could enhance the tourism potential of tourist villages.

Kusuma (2022) conducted community empowerment using the Enabling, Empowering, and Protecting paradigm, which involves a collaborative approach between the government and the community. This effort began by establishing regulations as a foundation for harnessing the potential of the community, thus creating opportunities for community empowerment as a focal point of growth. Sutawa (2012) found that community empowerment is crucial for tourism growth. The reason for community empowerment and involvement in tourism development is to bring advantages to both the community of the tourist destination and the community's efforts to preserve its culture and nature, thereby achieving sustainable tourism.

Researchers from the Cultural Heritage Preservation Research Institute concluded that the government needed to effectively form and implement laws, policies, and regulations. These laws, policies, and regulations should create a friendly environment for strengthening sustainable tourism in Lake Toba. Collaboration with the private sector and contributions from local communities can play a role in introducing, strengthening, and maintaining sustainable tourism (Anele, 2020). Anele and Sam-Otuonye (2021) found that scholars' recommendations significantly strengthened sustainable tourism in Lake Toba.

Hajar et al. (2021) conducted research that grouped the tourism potential owned by regions in the Lake Toba area. They suggested that these areas can be developed and managed well according to their potential and can create superior and competitive tourism. Through a SWOT analysis, Rambe et al. (2022) found that local tourism could be improved to support the development of Lake Toba tourism.

Sustainable Ecotourism

The World Tourism Organization defines sustainable tourism as tourism that considers both its present and future economic, social, and environmental effects. It aims to meet the demands of visitors, industry, the environment, and host communities (United Nations Department of Economic and Social Affairs Sustainable Development, 2024). Sustainable tourism encompasses economic, environmental, and sociocultural factors. Nature is the primary tourism resource. Destruction of nature poses a significant threat to sustainable tourism. Therefore, tourism development must prioritize the inclusion of the environment as a crucial resource that needs to be enhanced in terms of quality and diversity.

Figure 3. Group photo of meeting delegates on the shores of Lake Toba.

From an economic perspective, tourism has emerged as a significant revenue generator for nations. However, tourism must positively impact the livelihoods of local residents in a given location, including aiding in poverty alleviation. From a sociocultural perspective, tourism fails to prevent the negative modification of culture while also contributing to heritage conservation and the development of local people's capabilities. Summary: Sustainable tourism refers to responsible tourism that aligns with Agenda 21. This includes tourism activities that support the conservation of nature and heritage, promote the development of a more sustainable industry, and contribute to public awareness and capacity-building. This definition was provided by the United Nations Environment Program (2002).

The Universitas Sumatera Utara designed community empowerment activities in tourism areas in two districts with Lake Toba beaches: Samosir Regency and Karo Regency. This activity aimed to implement a community empowerment program through the International Community Service Program. The program was attended by several hundred people from 38 State Universities within the scope of the State Universities Cooperation Agency for the Western Region of Indonesia, including both local and international students. As a follow-up to this program, a meeting was held between the leaders of member universities and the leaders of the Samosir and Karo Regency governments. The atmosphere of the meetings between the two parties is shown in Figure 3.

Discussion

Empowerment is the systematic and strategic process of enhancing the capabilities, authority, and resilience of marginalized individuals or groups, enabling them to effectively challenge and resist oppressive forces across various domains and sectors of life (Eko, 2002). Local community empowerment can be understood from two perspectives. First, empowerment was observed within the framework of establishing the community's position of influence. The community's position is not that of passive recipients reliant on external entities such as the government for gifts. Instead, they are active actors who act independently and participate in decision-making. Engaging in autonomous action does not imply evading government obligations. It is indisputably the responsibility of the state to offer public services, such as healthcare, education, housing, and transportation to communities. Being an independent society participant entails the freedom and ability to explore and cultivate one's potential, manage their surroundings and resources, autonomously address challenges, and actively engage in shaping the political landscape inside the state. Individuals engage in development and governance activities (Eko, 2002).

Community empowerment can be defined as a localized process that focuses on enhancing the capabilities and self-determination of local communities. This approach is based on the principles of democratic involvement and critical reflection by members, enabling them to actively contribute to decision-making regarding circumstances and resources.

Developing a tourism business requires prioritizing community empowerment to guarantee that local communities receive benefits (Adebayo and Butcher, 2023). The Tourism Office has the potential to promote empowerment within the community and should be encouraged to further enhance its role in addressing the issue of insufficient community empowerment. Tourism development should prioritize the preservation of the environment and local culture in destination locations while also actively contributing to initiatives aimed at conserving natural resources and safeguarding cultural heritage. Tourism development should function as an industrial sector capable of enhancing the standard of living for local residents by decreasing the poverty rate in the region (Luo and Bao, 2019).

Assessment of Sustainable Tourism Supported by Community Empowerment

A questionnaire was distributed among the community to assess the importance of treatment in tourism management.

There were 20 questions categorized into three components: local communities, community empowerment, and government. Each component was divided into several sub-components.

The community component consists of the following seven sub-components:

1. Local governments integrate sustainability programs in tourism development.
2. Local communities are involved in planning and managing tourism destinations.
3. Training programs are in place to improve community skills in the tourism sector.
4. Tourism infrastructure supports environmental sustainability.
5. Tourism programs promote and maintain local cultural diversity.
6. There are economic incentives for local communities to participate in tourism projects.
7. Tourism programs ensure gender equity in community empowerment.

The empowerment component includes:

8. Tourism programs ensure active community participation in destination promotion.
9. Local communities have adequate access to information about the tourism industry.
10. Community empowerment in tourism takes into account social welfare aspects.
11. Local communities have a say in decisions that affect their tourism destination.
12. The government provides incentives to small and local businesses in the tourism sector.
13. Tourism programs focus on reducing negative impacts on the environment.
14. Local communities are empowered to promote the uniqueness and authenticity of tourism destinations.
15. Monitoring and enforcement mechanisms are in place against tourism practices that harm local communities.

The next component is government, which is divided into five sub-components:

16. The government provides funding for community initiatives in sustainable tourism development.
17. Sustainable tourism is measured not only in economic terms but also socially and environmentally.
18. Local communities are given the opportunity to maintain their traditions and lifestyles in tourism development.
19. There are educational programs for tourists on the importance of respecting local culture and environment.
20. Local communities feel that sustainable tourism provides tangible benefits to their lives.

The questionnaire was distributed to the tourism user community using a four-point Likert scale. A total of 80 respondents answered these questions. The average results of the responses are presented in Table 1 and graphically shown in Figure 4.

Table 1. Question Response

No.	Questions	Responses	Subcomponents
Community			
1.	Local communities are empowered to promote the uniqueness and authenticity of tourist destinations.	3,75	Promotion
2.	Local communities are given opportunities to maintain their traditions and lifestyles in tourism development.	3,75	Tradition and Lifestyle
3.	Local communities are involved in planning and managing tourism destinations.	3,725	Engagement
4.	Local communities feel that sustainable tourism provides tangible benefits to their lives.	3,65	Benefits
5.	There are economic incentives for local communities to participate in tourism projects.	3,6	Incentives
6.	Local communities have adequate access to information about the tourism industry.	3,55	Access
7.	Local communities have a say in decisions that affect their tourism destination.	3,5125	Take part in Decisions
Empowerment			
8.	Tourism programs promote and maintain local cultural diversity.	3,725	Promotion
9.	Community empowerment in tourism takes into account social welfare aspects.	3,725	Wellbeing
10.	Sustainable tourism is measured not only in economic terms but also socially and environmentally.	3,725	Environmental and Social
11.	There are training programs to improve community skills in the tourism sector.	3,65	Program
12.	There are educational programs for tourists about the importance of respecting the local culture and environment.	3,6375	Education to Tourists
13.	There are educational programs for tourists on the importance of respecting local culture and environment.	3,6125	Environmental Impact
14.	The tourism program ensures active community participation in destination promotion.	3,525	Promotion
15.	The tourism program ensures gender equity in community empowerment.	3,4375	Gender
Government			
16.	Local governments integrate sustainability programs into tourism development.	3,575	Program Integration
17.	Tourism infrastructure supports environmental sustainability.	3,575	Environment
18.	There is a mechanism for monitoring and enforcing laws against tourism practices that harm local communities.	3,5375	Monitoring
19.	The government provides funding for community initiatives in sustainable tourism development.	3,4375	Fund
20.	The government provides incentives to small and local businesses in the tourism sector.	3,425	(MSME) Incentives

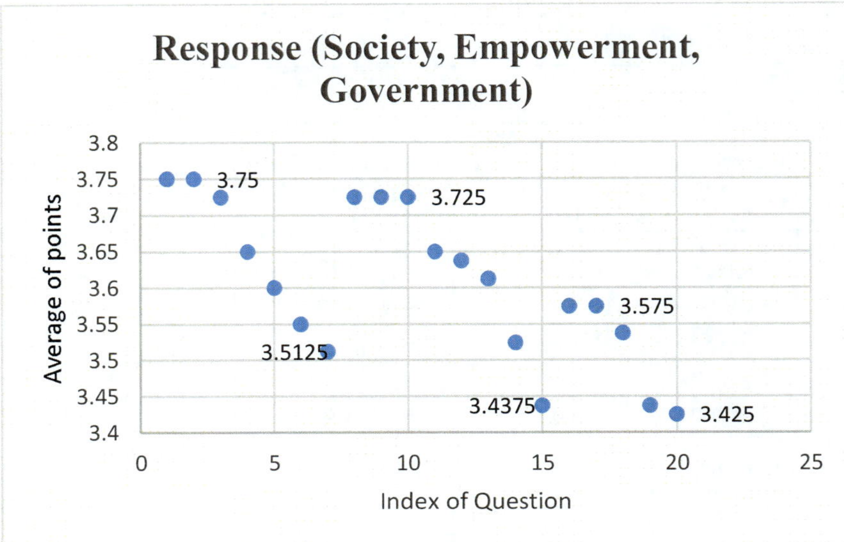

Figure 4. Response to the questionnaire.

From Table 1 and Figure 4, it can be seen that the community component has the highest score with an average score of 3.65, followed by the empowerment component as an action program with an average score of 3.63, and then the government component with an average score of 3.51. The community best achieves the promotion of local uniqueness and lifestyle. The empowerment action program also focuses on promoting attention to welfare, environmental, and social aspects. Alternatively, the government should integrate sustainability programs in tourism development as well as the environment. It can be concluded that tourism promotion must be conducted by the community and accompanied by community empowerment programs. The sustainability aspects of tourism development and the environment must be implemented by the relevant governments.

Conclusion

Community empowerment is important in improving the quality of potential tourism areas in North Sumatra, especially in major destinations such as Lake Toba. Community empowerment is defined as a process that focuses on the active involvement of local communities in tourism-related management and decision-making. Through the Community Based Tourism

(CBT) model, communities play a role in various tourism activities and benefit directly from this sector. The research also shows that community involvement helps maintain cultural and environmental sustainability. The government also has a key role in supporting the success of this program through regulations and policies that favor tourism sustainability. Collaboration between the government, community, and private sector is needed to realize sustainable tourism that benefits all parties. Community empowerment and collaboration with the government and the private sector are key to sustainable tourism management and have a positive impact on local welfare.

Disclaimer

None.

References

Adebayo, A. D., Butcher, J. 2023. Community Empowerment in Nigeria's Tourism Industry: An Analysis of Stakeholders' Perceptions. *Tourism Planning & Development.* (Vol. 20, No. 4, p. 583–603).

Anele, K. K, Sam-Otuonye, C. C. 2021. Sustainable Tourism: Evidence from Lake Toba in North Sumatra, Indonesia. *ASEAN Journal on Hospitality and Tourism*. (Vol. 19, No. 1, pp. 52-62).

Anele, K. K. 2020. An Assessment of Sustainable Tourism Development in North Sumatra. *Proceedings of the International Conference on Culture Heritage, Education, Sustainable Tourism, and Innovation Technologies*. (p. 207-218).

Basyuni, M., Putri, L. A. P., Nainggolan, B., Sihaloho, P. E. 2014. Growth and biomass in response to salinity and subsequent fresh water in mangrove seedlings Avicennia marina and Rhizophora stylosa. *Jurnal Manajemen Hutan Tropika* 20(1): 17-25.

Eko, S., 2002, Pemberdayaan Masyarakat Desa. *Materi Diklat Pemberdayaan Masyarakat Desa, yang diselenggarakan Badan Diklat Provinsi Kaltim, Samarinda.*

Ginanjar, R., Runingsawitri, H. S. 2023. Community Empowerment in Tourism Development: Concepts and Implications. *The Eastasouth Management and Business*. (Vol. 1, No. 3, pp. 111–119).

Hajar, S., Supriyono, B., Muluk, M. R. K., Said, A. 2021. Tourism Potential Planning Based Governance in the Lake Toba. (Vol. 11, No. 2, pp. 549-558).

Hsua, C. Y., Chenb, M. Y., Nyaupanec, G. P., Lind, S. H. 2020. Measuring Sustainable Tourism Attitude Scale (SUS-TAS) in an Eastern Island Context. *Tourism Management Perspectives*. (Vol. 33).

Khan, I. U. 2022. Residents' Satisfaction with Sustainable Tourism: The Moderating Role of Environmental Awareness. *Tourism Critiques: Practice and Theory*. 3(1): 72-87.

Kusuma, A. R. 2022. Community Empowerment Through Tourism Development in East Kalimantan. *Jurnal Ilmiah Ilmu Administrasi Publik: Jurnal Pemikiran dan Penelitian Administrasi Publik.* 12(2): 489-500.

Luo, X., Bao, J. 2019. Exploring the Impacts of Tourism on the Livelihoods of Local Poor: The Role of Local Government and Major Investors. *Journal of Sustainable Tourism*. 27(3): 344–359.

Purworini, D., Chasana, R. R. B., Palupi, P., Setyawan, S., Triyono, A., Yuwono, I. T. 2023. The Role of Community Empowerment in the Development of a Marketing Communication Strategy for Kembang Kuning Tourism Village. *ASEAN Journal of Community Engagement*. 7(1): 113–126.

Rachmawati, E., Hidayati, S., Rahayuningsih, T. 2021. Community Involvement and Social Empowerment in Tourism Development. *Media Konservasi* 26(3): 193-201.

Rambe, T., Simanjuntak, D. H. P., Gafari, M. O. F., Amal, B. K. 2022. The Development of Lake Toba Tourism by Strengthening Local Wisdom. *The Seybold Report*. 17(9): 200-211.

Renon, M., Peysson, J. 2020. Sustainable Tourism: A Quantitative Study on Awareness, Attitudes and Behaviors of Young Adults. Department of Business Administration Master's Program in Business Development and Internationalisation.

Sutawa, G. K. 2012. Issues on Bali Tourism Development and Community Empowerment to Support Sustainable Tourism Development. *Procedia Economics and Finance* 4: 413-422.

United Nations Department of Economic and Social Affairs Sustainable Development. 2024)(https://sdgs.un.org). Accesed on: 19th June 2024.

United Nations Environment Program. UNEP *Annual Report.* 2002. www.unep.org/publications/contents/pub_details_search.asp?ID=274. Accesed on: 19th June 2024.

Chapter 17

Characteristics and the Potential of North Sumatra Plants as Raw Materials for Particleboard

Apri Heri Iswanto[1,*]
Halimatuddahliana[2]
and Hamidah Harahap[2]
[1]Faculty of Forestry, Universitas Sumatera Utara, Medan, Indonesia
[2]Faculty of Engineering, Universitas Sumatera Utara, Medan, Indonesia

Abstract

On the island of Sumatra, the forested areas in North Sumatra contain several species of local plants that are of high value and can be developed as alternative raw materials for composite panels. Plant species from North Sumatra that are potential raw materials for composite panels include kemenyan and raru plants. Based on the results of species identification of frankincense trees, four kemenyan species were found in North Sumatra: toba (*Styrax sumatrana*), durame (*Styrax benzoin*), bulu (*Styrax paralleloneurum*), and dairi (*Styrax sp*). Meanwhile, three species of raru plants were found: dahanon (*Cotylelobium lanceolatum*), *pulut* (*Cotylelobium melanoxylon*), and dan songal (*Vatica pauciflora*). The physical properties of the wood showed that the three types of incense wood had medium specific gravity values. These results fulfil the ideal compression ratio of lignocellulosic materials for composite board raw materials, in this case, particle board. However, raru wood has a high specific gravity and

* Corresponding Author's Email: apri@usu.ac.id.

In: Natural Resources, Environment, and Food and Energy Security …
Editors: Mohammad Basyuni and Severino Salmo III
ISBN: 979-8-89530-283-5

thus has the potential to increase the strength of particleboard through hybrid particle board products.

Keywords: plants from North Sumatra, basic properties of wood, particleboard

Introduction

Indonesia has approximately 4,000 tree species that can potentially be used as timber for construction. However, only approximately 400 species (10%) have economic value, and even fewer, 260 species, have been classified as commercial timber (Soerianegara and Lemens, 2002). With the development of the forestry sector, which has been highly exploitative and led to an increase in the rate of forest destruction, existing timber species are likely to have become scarce. This has impacted the forestry industry, which uses wood as its main raw material. Many forestry industries have gone out of business because of difficulties in obtaining raw materials. Some surviving industries, such as pulp and paper, rely on industrial forests to supply raw materials. In addition to industrial plantations and community forests, the particleboard industry can use waste or unused wood waste to meet its raw material requirements.

According to statistics from the Ministry of Environment and Forestry (2022), North Sumatra has a forest area of approximately three million hectares and produces several local wood species that can be studied for their properties and characteristics for appropriate use. Among the existing tree species, kemenyan and raru grow well in North Sumatra. Research has shown that four species of kemenyan trees grow in North Sumatra: kemenyan toba (*Styrax sumatrana*), durame (*Styrax benzoin*), bulu (*Styarax paralleloneurum*) and dairi (*Styrax* sp). Three species were found among raru plants: raru dahanon (*Cotylelobium lanceolatum*), raru pulut (*Cotylelobium melanoxylon*) and raru songal (*Vatica pauciflora*). Characterisation of their physical and mechanical properties indicated that these six North Sumatran timber species have potential as construction materials, furniture, and raw materials for the production of composite panels for laminated wood, oriented strand board and particleboard.

Discussion

Kemenyan Wood

In addition to producing sap, kemenyan trees can also produce wood, particularly from unproductive trees. Kemenyan wood is of good quality, with strength classes III–IV, and good dimensional stability (Iswanto et al. 2016). Owing to these properties, kemenyan wood can be used for furniture, handicrafts, and lightweight construction.

Table 1. Physical and mechanical properties of three kemenyan wood species

Wood Test Parameter	Kemenyan Wood Species		
	S. benzoine	*S. paralleloneurum*	*S. sumatrana*
Physical properties			
Specific gravity:			
Green	0.48	0.43	0.54
Air dry	0.53	0.47	0.59
Oven dry	0.56	0.48	0.62
Shrinkage (%):			
Longitudinal	1.54	1.03	1.64
Radial	6.65	4.63	6.26
Tangential	6.73	5.20	5.58
Volume	14.30	10.55	12.95
Mechanical properties			
MOR (kg/cm^2)	770.45	637.52	729.82
MOE (kg/cm^2)	85,900	77,685	77,649
Hardness (kg)	507	391	599
Compression (kg/cm^2)	355	313	351

Source: Iswanto et al. (2016).

Specific Gravity of Kemenyan Wood

Bowyer et al. (2003) stated that density is the mass or weight per unit volume of wood expressed in kilograms per cubic meter or pounds per cubic foot. Iswanto et al. (2016) reported that three species of kemenyan wood, namely bulu (*Styrax paralleloneurum*), durame (*Styrax benzoin* Dryan), and toba (*Styrax sumatrana* J. J. Sm), produced wood with specific gravity values ranging from 0.43 to 0.62 (Table 1). In this study, kemenyan tobacco wood was found to have the highest specific gravity. However, kemenyan wood still has the same range of specific gravity values as several other tropical wood species, including *Sapium baccatum*, *Macaranga gigantea*,

Endospermum malaccense, *Hevea brasiliensis*, and *Eucalyptus grandis*, which have a specific gravity range of 0.4–0.5 (Hamdan et al. 2020).

During the manufacturing of composite panels, low to medium density wood is the preferred raw material for panel production. This is because boards with this density have a better compression ratio and can therefore produce better board strength. Some studies, such as those by Bufalino et al. (2012), Maulana et al. (2020), and Lee et al. (2015), have reported that the compression ratio is one of the factors that determines the quality of particleboard products.

Kemenyan Wood Shrinkage

Similar to wood, the kemenyan wood also undergoes shrinkage. Furthermore, Iswanto et al. (2016) reported that the shrinkage of kemenyan wood varies. Based on Table 1, the kemenyan wood that experienced the highest shrinkage was kemenyan bulu. The high shrinkage value found in kemenyan bulu indicates that the internal stress in this species of wood is high and has the potential to cause defects during wood processing. In contrast, kemenyan durame is known to be the kemenyan species with the lowest shrinkage value. The shrinkage that occurs is influenced by the water content and wood density (Lima et al. 2014).

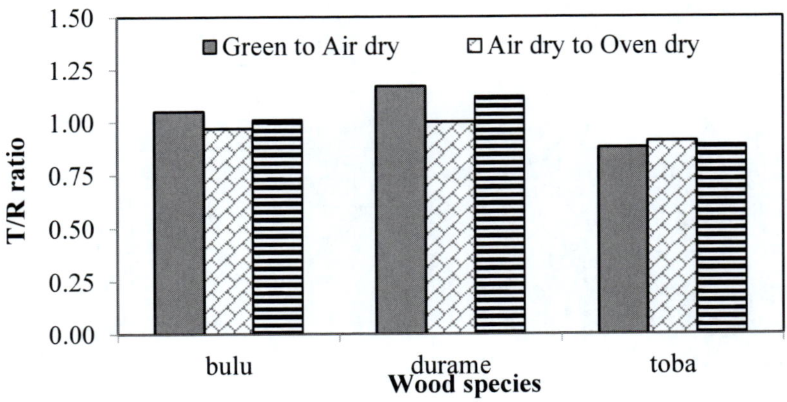

Figure 1. T/R ratios of three kemenyan wood species.

The percentages of wood shrinkage in the longitudinal, radial, and tangential directions were 1.03–1.64%, 4.63–6.65%, and 5.20–6.73%, respectively. The obtained volume shrinkages were in the range of 10.55–14.30%. As stated by the Forest Product Laboratory (1999), the swelling and shrinkage of wood are not equal in each direction along the main axis. Tangential wood shrinkage was greater than radial wood shrinkage because the arrangement of the radius extended radially, thus limiting shrinkage in the radial plane. Another cause is the thinness of the cell wall and the greater number of noctahs in the radial plane (Brown et al. 1952).

The tangential to radial shrinkage (T/R) ratio values of the three incense woods ranged from 0.88 to 1.17, as shown in Figure 1. The T/R ratios indicated that the three kemenyan wood species had good dimensional stability. According to Phansin and de Zeeuw (1980), wood has good dimensional stability if the T/R ratio is 1.00.

Modulus of Elasticity (MOE) and Modulus of Rupture (MOR) of Kemenyan Wood

The MOE and MOR values of the three types of frankincense wood are listed in Table 1. As reported by Iswanto et al. (2016), bulu frankincense wood has higher MOE and MOR values than Durame and Toba kemenyan wood species, which range from 77,685 to 85,900 kg/cm^2 and 637 to 770 kg/cm^2, respectively. The MOE and MOR values of these three kemenyan wood species are equivalent to those of several tropical wood species, such as *Paraserianthes falcataria*, *Macaranga gigantea*, *Hevea brasiliensis*, *Eucalyptus grandis*, which have MOE values ranging from 77,000 to 85,000 kg/cm^2 respectively. The MOR values are also equivalent to those of *Sapium baccatum*, *Macaranga gigantea,* and *Endospermum malaccense* (Hamdan et al. 2020).

Hardness and Compression Strength of Kemenyan Wood

Wood hardness can be interpreted as the ability of wood to resist damage and scratching on its surface (Mardikanto et al. 2011). Therefore, wood hardness is closely related to its density. Wood hardness is one of the factors considered in the analysis of wood designation. This is related to the intended use of the wood in terms of safety. In addition to wood density, wood hardness is also influenced by several factors, such as fibre size, fibre arrangement, and the binding strength of wood fibres.

Iswanto et al. (2016) reported that the parallel compression and hardness values of kemenyan wood were in the ranges of 313–355 kg/cm^2 and 391–599 kg/cm^2, respectively. The effect of the wood-specific gravity caused kemenyan bulu to have the lowest specific gravity among the kemenyan species.

Chemical Components of Kemenyan Wood

Analysis of the chemical components of wood plays an important role in determining the contents of hemicellulose, lignin, cellulose, and extractives. According to Fengel and Wegener (1995), wood contains various major chemical components (cellulose, hemicellulose, and lignin) and minor components (minerals and extractives).

The chemical properties of wood affect its durability, colour, strength, and processing, and the chemical properties of kemenyan wood are important to consider for its use in various applications. Iswanto et al. (2019) reported the chemical content of kemenyan toba wood from two different natural habitats, North Tapanuli and the Pakpak Bharat Regency, as shown in Table 2.

Table 2. Chemical components of kemenyan toba (*S. sumatrana*) from North Tapanuli and Pakpak Bharat Regency

Location	Lignin	Holocellulose	α-Cellulose	Hemicellulose	Extractives in Ethanol Benzene
North Tapanuli	38.5	50.92	25.05	25.87	10.58
Pakpak Bharat	24.77	69.8	39.87	29.92	2.42

Source: Iswanto et al. (2019). All values are in % units.

The lignin content of kemenyan toba wood from North Tapanuli was higher than that of wood from Pakpak Bharat because of the hilly topography of Pakpak Bharat. This is supported by the findings of Stout et al. (2014), who reported in their study that the lignin content increased slightly in mountainous areas. In wood cell walls, lignin acts as a binder between cellulose and hemicellulose. Lignin aids in the gluing process during the manufacturing of composite panels. Several studies have investigated the use of lignin as an adhesive material, including those by Gadhave et al. (2019), Yang et al. (2023), Dorn et al. (2023), Meng et al. (2023), and Aristri et al. (2021).

A striking difference between the two types of wood from different growing areas can also be observed in their extractive content. The wood from North Tapanuli had a higher extractive content. Extractives are closely related to the natural durability of wood. A high extractive content gives wood better durability. In addition to its positive effect on the durability of wood, the extractive content adversely affects the production process of the composite board. The presence of extractives hinders the penetration of the adhesive into the wood particles, and even during the hot-pressing process, the extractives can cause blowing. In principle, the presence of extractives affects the performance of the adhesive in the hot forging process of particleboard or other composite products, as reported by several researchers, including Santoso et al. (2016), Hashim et al. (2011), and Nussbaum et al. (2002).

Natural Durability of Kemenyan Wood

The durability of wood is closely related to its chemical composition. Therefore, extractives play an important role in the natural durability of wood. Indonesia is a tropical country containing a diverse range of destructive organisms. Termites are among the most common destructive organisms. Termites use wood for consumption and breeding (Awadzi et al. 2004) and are often found to destroy various cellulose-containing objects, such as furniture, house poles, and books.

Natural preservatives found in wood can help protect it from attack by these insects. Kemenyan wood, which is often used by communities for various purposes, has considerable potential for attack by destructive organisms. Iswanto et al. (2021) analysed the durability of kemenyan toba wood against destructive organisms by conducting grave tests on kemenyan toba wood obtained from different growing sites. Iswanto et al. (2021) reported that kemenyan toba wood from Pakpak Bharat had lower resistance to subterranean termite attack than kemenyan toba wood from North Tapanuli, as indicated by the higher weight loss values for kemenyan toba wood from Pakpak Bharat (Figure 2).

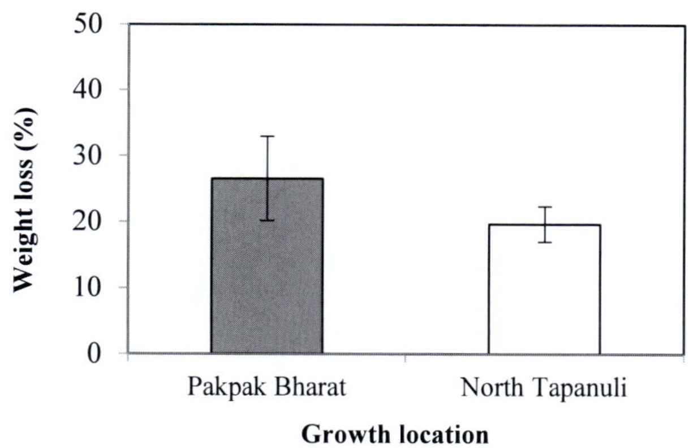

Figure 2. Weight loss values of kemenyan toba wood from two different growth sites.

These differences are thought to be due to the levels of chemical components such as cellulose and extractive substances in the two test samples. The results of the study conducted by Iswanto (2019) showed that the cellulose content was lower in the kemenyan toba wood from North Tapanuli, while the extractive substance content was higher. Cellulose, a food source for termites (Scheffrahn, 1991), is present in small amounts in wood samples from North Tapanuli, and extractives that are toxic (Kadir, 2017) cause termites to prefer wood samples from Pakpak Bharat. Although kemenyan wood from North Tapanuli is more durable than that from Pakpak Bharat, the kemenyan wood from both locations is classified as durability class V.

Raru Wood

The Bataknese use raru wood as an ingredient in the traditional drink 'Tuak.' Three species of raru wood are found in North Sumatra: dahanon (*Cotylelobium lanceolatum*), sonal (*Vatica pauciflora*), and pulut (*Cotylelobium melanoxylon*). In terms of their physical and mechanical properties, the three raru wood species listed in Table 3 all fall into strength classes I and II. This indicates that this wood has great potential as a construction material.

Specific Gravity of Raru Wood

Based on Table 3, the three types of raru wood have a relatively high specific gravity, which is above 0.8 and even close to 1. The specific gravity of raru wood is equivalent to that of gelam wood (*Maleleuca sp*) and close to that of ulin wood (*Eusideroxylon zwageri*), as reported by Mawadah and Basyaruddin (2021). This high specific gravity indicates that raru wood has high strength and is suitable for use as a structural component in buildings. According to the classification of wood strength classes based on specific gravity, raru wood is classified into strength classes I and II. During the manufacturing of composite panels, particularly particleboard, this wood is less suitable as a primary raw material because of its high specific gravity; however, this wood is a suitable facing material for hybrid particleboard or sandwich particleboard. Typically, a high-strength material is placed on the surface to withstand loads that are in direct contact with it. In hybrid particle board products, raru wood is used as the outer layer (surface) to increase the board strength. The problem with particleboard, in general, is its lower strength compared to wood. Hybrid technology using high-strength materials is expected to overcome this problem. Several studies on increasing the strength through hybrid particleboard have been conducted by Iswanto and Ompusunggu (2019), Iswanto et al. (2020), Fueangvivat et al. (2014), and Silva et al. (2018).

Table 3. Physical–mechanical properties of *raru wood*

Wood Test Parameter	Raru Wood Species		
	Pulut Wood* (*C. melanoxylon*)	Dahanon Wood** (*C. lanceolatum*)	Songal Wood** (*Vatica pauciflora*)
Physical Properties			
Specific gravity:			
Green	0.82	0.86	0.83
Air dry	0.91	0.83	0.89
Oven dry	0.99	0.89	0.96
Shrinkage (%):			
Longitudinal	1.75	0.35	0.27
Radial	4.51	1.54	1.7
Tangential	4.17	1.61	2.27
Volume	7.95	3.46	4.18
T/R ratio	0.92	1.05	1.36
Mechanical Properties			
MOR (kg/cm^2)	1,847	1,500	1,558
MOE (kg/cm^2)	270,569	213,237	157,465
Compression (kg/cm^2)	736	503	632

Source: *Putra (2019), **Khairani (2021).

Shrinkage of Raru Wood

Wood shrinkage is the reduction in wood dimensions corresponding to a reduction in water content below the fibre saturation point (Tsoumis, 1991). According to Haygreen (1989), the change in wood dimensions is not influenced by the loss of free water from the cell cavity but by the loss of bound water from the cell wall, which causes the cell and wood to shrink. Based on Table 3, among the three types of raru wood, the largest shrinkage value occurred in raru pulut (*C. melanoxylon*) wood. This is based on the findings of Sekha et al. (1967) and Schroeder (1972) that the shrinkage value of wood volume is due to high specific gravity.

The tangential to radial shrinkage ratio (T/R) for *C. melanoxylon* wood was close to one and superior to those of other species. This indicates that raru wood has good dimensional stability because the difference in the tangential shrinkage of Raru wood is not too large compared to the radial shrinkage. This statement is also supported by the study of Panshin (1980), who found that when close to one, the T/R value is more stable.

MOE, MOR, and Compression of Raru Wood

The highest MOE, MOR, and fibre parallel compression values were obtained for the *C. melanoxylon* raru wood. This is consistent with the specific gravity value of this species being relatively higher than those of the other two species. The strength of wood can determine the strength class in which it is used. Based on the results of the MOE and compression strength parallel to the grain of raru wood, raru wood is a strength class I wood, which has a minimum MOE and compression strength parallel to the grain of 125,000 kg/cm^2 and 130 kg/cm^2, respectively.

Chemical Components of Raru Wood

The main chemical composition (lignin, holocellulose, α-cellulose, hemicellulose, and extractives in ethanol-benzene) of the three species of raru wood, namely *C. melanoxylon*, *C. lanceolatum*, and *Vatica pauciflora* have variations in each chemical component that may affect their characteristics (Table 4). Lignin is one of the main chemical components in plants. Lignin accumulates in the stems of plants and acts as a component that binds other components together to allow trees to stand upright, usually likened to cement in concrete (Setiati et al. 2016). The lignin contents of the three raru wood species were in the range of 24–30%, with *C. melanoxylon* wood having the lowest lignin content. These three species of raru wood have the same range of lignin content as several other tropical wood species, including *Duabanga grandiflora*, *Toona ciliata*, *Anthocephalus chinensis*, *Gmelina arborea*, and *Callicarpa poilanei*, which have lignin contents in the range of 20–27% (Wu et al. 1990) and *Dacryodes edulis*, *Strombosia grandifolia*, *Hibiscus tiliaceus*, and *Terminalia superba*, which have lignin contents in the range of 29–31% (Gerard et al. 2019). Lignin acts as a binder in cell walls. In the walls of buildings, lignin acts as an adhesive for sand and stone. Therefore, during the manufacturing of composites such as particleboard, lignin plays a role in supporting the bonding process in addition to the adhesive strength provided by synthetic adhesives.

The cellulose contents of the three raru wood species were in the range of 41–42%. Compared to some other tropical wood species such as *Dacryodes edulis*, *Octoknema sp.*, *Strombosia grandifolia*, and *Hibiscus tiliaceus*, which have cellulose contents ranging from 44 to 47% (Gerard et al. 2019), the cellulose content of raru wood is slightly lower. As a component of the cell wall, cellulose plays a role in the formation of cell walls and plant fibres. Cellulose is a brick that constitutes the main component of the cell wall. The presence of cellulose is the main requirement for particleboard manufacturing. As defined by Maloney (1993), particleboard is a composite product made from wood or non-wood lignocellulosic materials bonded together with adhesives and hot-pressed.

Table 4. Chemical components of three species of raru wood

Wood species	Lignin	Holocellulose	α-Cellulose	Hemicellulose	Extractives in Ethanol Benzene
C. melanoxylon	24.19	61.62	41.04	20.6	14.1
C. lanceolatum	28.69	61.3	41.89	23.2	10.6
Vatica pauciflora	30.12	66.2	42.95	13.40	11

Source: Iswanto et al. (2021).

The extractive contents of the three raru wood species were in the range of 10–14%. Compared with several other tropical wood species, such as *Albizia lebbeck*, *Anthocephalus cinensis*, and *Litsea roxburghii*, as reported by Wibisono et al. (2018), the three raru wood species had higher extractive content values. In particleboard production, the presence of extractives is an inhibiting factor in the gluing process; therefore, these extractives, especially in high-content wood, must first be eliminated through various particle pretreatments such as hot and cold-water immersion and autoclaving or steam treatment, as has been performed by several researchers, including Iswanto et al. (2018), Iswanto and Togatorop (2019), Bekta et al. (2013), and Iewkittayakorn et al. (2017).

Conclusion

An evaluation of the potential of local plant species of North Sumatra was conducted, specifically frankincense (*Styrax* spp.) and raru (*Cotylelobium* spp. and *Vatica pauciflora*), as alternative raw materials for particleboard. From the analysis of physical and mechanical characteristics, frankincense showed a medium specific gravity, which is ideal for compression ratios in particleboard production, while raru, with a high specific gravity, has the potential to increase particleboard strength in hybrid products. Frankincense has good dimensional stability and mechanical properties suitable for use in the furniture and light construction industries. On the other hand, raru, which belongs to strength classes I and II, exhibits high strength and dimensional stability, making it suitable as a structural material in construction applications. The combination of frankincense and raru in the production of hybrid particleboard can improve the overall quality of the product, especially in terms of strength and durability.

Disclaimer

None.

References

Aristri, M. A., Lubis, M. A. R., Laksana, R. P. B., Falah, F., Fatriasari, W., Ismayati, M., Wulandari, A. P., Nurindah, Ridho, M. R. 2021. Bio-Polyurethane Resins Derived from Liquid Fractions of Lignin for the Modification of Ramie Fibers. *Jurnal Sylva Lestari* 9(2): 223-238.

Awadzi, T. W., Cobblah, M. A., Madsen, H. B. 2004. The role of termites in soil formation in the tropical semi-deciduous forest zone, Ghana. *Danish J Geogr* 104 (2): 27-34.

Bekhta, P., Korkut, S., Hiziroglu, S. 2013. Effect of Pretreatment of Raw Material on Properties of Particleboard Panels Made from Wheat Straw. *Bioresources* 8: 4766-4774. 10.15376/biores.8.3.4766-4774.

Bowyer, J. L., Shmulsky, Haygreen, J. G. Forest Products and Wood Science - An Introduction. Fourth edition. Iowa (US): Iowa State University Pr; 2003.

Brown, H. P., Panshin, A. J., Forsaith, C. C. *Text Book of Wood Technology* Vol. II. New york: McGraw-Hill Book Company Inc; 1952.

Bufalino, L., Albino, V. C. S, de Sá, V. A., Corrêa, A. A. R., Mendes, L. M, Almeida, N. A. 2012. Particleboards Made from Australian Red Cedar: Processing Variables and Evaluation of Mixed-Species. *Journal Tropical Forest Science* 24(2):162-172.

Dorn, L., Thirion, A., Ghorbani, M., Olaechea, L. M., Mayer, I. 2023. Exploring Fully Biobased Adhesives: Sustainable Kraft Lignin and 5-HMF Adhesive for Particleboards. *Polymers* 15, 2668. https://doi.org/10.3390/polym15122668.

Fengel, D., Wegener, G. KAYU: Kimia, Ultrastuktur, Reaksi-reaksi. Gadjah Mada University Press. Yogyakarta; 1995.

Foerst Product Laboratory. *Wood Hand Book: Wood as an Engineering Material. Agric Handbook 72*. Washington DC. US department; 1999.

Fueangvivat, V., Jarusombuti, S., Hiziroglu, S., Bauchongkol, P. 2014. Properties of Sandwich-Type Particleboard Panels Made from Rubberwood and Eastern Redcedar. *Forest Products Journal* 64 (1-2): 72–76. doi: https://doi.org/10.13073/FPJ-D-13-00082.

Gadhave, R., Srivastava, S., Mahanwar, P., Gadekar, P. 2019. Lignin: Renewable Raw Material for Adhesive. *Open Journal of Polymer Chemistry* 9, 27-38. doi: 10.4236/ojpchem.2019.92003.

Gerard, J., Paradis, S., Thibaut, B. 2019. Survey on the chemical composition of several tropical wood species. *Bois et For.ts des Tropiques* 342: 79-91. Doi: 10.19182/bft2019.342.a31809.

Hamdan, H., Nordahlia, A. S., Anwar, U. M. K, Mohd Iskandar, M., Mohamad Omar, M. K, Tumirah, K. 2020. Anatomical, physical, and mechanical properties of four pioneer species in Malaysia. *J Wood Sci* 66 (59): 1-9. doi: 10.1186/s10086-020-01905-z.

Hashim, R., Amini, M. H. M., Sulaiman, O. 2011. Some properties of particleboard panels treated with extractives of Cerbera odollam tree. *Journal of Composite Materials* 45(8):901-906. doi:10.1177/0021998310377941.

Haygreen, J. G., Bowyer, J. L. *Forest Products and Wood Science*, Iowa State University Press, USA; (1989).

Iewkittayakorn, J., Khunthongkaew, P., Chotigeat, W., Sudesh, K. 2017. Effect of Microwave Pretreatment on the Properties of Particleboard Made from Para Rubber Wood Sawdust with the Addition of Polyhydroxyalkanoates. *Sains Malaysiana* 46(9): 1361–1367.

Iswanto, A. H, Susilowati, A., Azhar, I., Riswan, Supriyanto, Tarigan, J. E., Fatriasari, W. 2016. Physical and Mechanical Properties of Local Styrax Woods from North Tapanuli in Indonesia. *J. Korean Wood Sci. Technol* 44(4): 539-550.

Iswanto, A. H., Siregar, Y. S., Susilowati, A., Darwis, A., Hartono, R., Wirjosentono, B., Rachmat, H. H., Hidayat, A., Fatriasari, W. 2019. Short Communication: Variation in chemical constituent of *Styrax sumatrana* wood growing at different cultivation site in North Sumatra, Indonesia. *Biodiversitas* 20(2): 448-452.

Iswanto, A. H., Tambunan, J., Susilowati, A., Hartono, R., Darwis, A. 2021. Short Communication: The resistance of Styrax sumatrana wood of varying growth sites and stem axial positions to subterranean termite (Coptotermes curvignathus) attack. *Biodiversitas* 22(6): 3192-3198.

Iswanto, A. H., Ompusunggu, P. 2019. Sandwich Particleboard (SPb): effect of particle length on the quality of board. *IOP Conference Series: Earth and Environmental Science* 374. 012002. 10.1088/1755-1315/374/1/012002.

Iswanto, A. H., Hakim, A. R, Azhar, I., Wirjosentono, B., Prabuningrum, D. S. 2020. The Physical, Mechanical, and Sound Absorption Properties of Sandwich Particleboard (SPb). *J. Korean Wood Sci. Technol* 48(1): 32-40.

Iswanto, A. H., Togatorop, R. M. M. 2019. Effect of Particle Immersing in Various Concentrations of Acetic Acid to Enhance the Sandwich Particleboard (SPb). in *International Conference on Basic Sciences and Its Applications, KnE Engineering* 20–29. doi 10.18502/keg.v1i2.4428.

Iswanto, A. H., Febrianto, F., Hadi, Y. S., Ruhendi, S., Hermawan, D., Fatriasari, W. 2018. Effect of Particle Pre-Treatment on Properties of Jatropha Fruit Hulls Particleboard. *J. Korean Wood Sci. Technol* 46(2):155-165.

Khairani, D. *Sifat Fisis dan Mekanis Kayu Raru Jenis Dahanon dan Songal. Undergraduated Thesis*, Universitas Sumatera Utara, Medan; 2021.

Lee, S. H., Ashaari, Z., Lum, W. C., Ng, P., San, Peng, T., Jinn, M., Chai, W., Chin, K. L. 2015. Properties of Particleboard with Oil Palm Trunk as Core Layer in Comparison to Three-Layer Rubberwood Particleboard. *Journal of Oil Palm Research* 27 (1): 67-74.

Lima, I. L., Longui, E. L., Freitas, M. L. M., Zanatto, A. C. S., Zanata, M., Florsheim, S. M. B., Bortoletto, Jr. G. 2014. Physical-Mechanical and Anatomical Characterization in 26-Year-Old Eucalyptus resinifera Wood. *Floresta e Ambiente* 21(1): 91-98.

Maloney, T. M. *Modern Particleboard and Dry - Process Fiberboard Manufacturing. Updated Edition.* San Francisco (US): Miller Freeman Pr; 1993.

Maulana, S., Sumardi, I., Wistara, N. J., Kim, N. H., Febrianto, F. 2020. Effects of compression ratio on physical and mechanical properties of bambooriented strand board. *IOP Conf. Series: Materials Science and Engineering* 935 012063 doi:10.1088/1757-899X/935/1/012063.

Mardikanto, T. R., Karlinasari, L., Bahtiar, E. T. *Sifat Mekanis Kayu*. Bogor (ID): IPB Pr; 2011.

Mawadah, F. Z., Basyaruddin. 2021. Pengaruh Penanaman Kayu Ulin dan Kayu Gelam Dalam Mortar Sebagai Balok Struktur Rumah Kayu Sederhana Berdasarkan Sifat Fisik dan Sifat Mekanik. *Rekayasa Sipil* 15(2): 95-103.

Meng, T., Ding, Y., Liu, Y., Xu, L., Mao, Y., Gelfond, J., Li, S., Li, Z., Salipante, P. F., Kim, H., Zhu, J. Y., Pan, X., Hu, L. 2023. In Situ Lignin Adhesion for High-Performance Bamboo Composites. *Nano Letters* 23(18): 8411-8418. https://doi.org/ 10.1021/acs.nanolett.3c01497.

Nussbaum, R., Magdalena, S. 2002. The effect of wood extractive content on glue adhesion and surface wettability of wood. *Wood and fiber science: journal of the Society of Wood Science and Technology* 34(1): 57.

Panshin, A. J., & de Zeeuw, C. *Text Book of Wood Technology. Structure, Identification, Properties and Uses of The Commercial Woods of The United States and Canada.* New york: McGraw-Hill Book Company, Inc; 1980.

Putra, A. R. Sifat *Fisis dan Mekanis Kayu Raru* (*Cotylelobium Melanoxylon*) Undergraduated Thesis, Universitas Sumatera Utara, Medan; 2019.

Santoso, M., Widyorini, R., Prayitno, T. A., Sulistyo, J. 2016. The Effects of Extractives Substances for Bonding Performance of Three Natural Binder on Nipa Fronds Particleboard *KnE Life Sciences* doi:10.18502/kls.v4i11.3868.

Sekha, A. C., Rajput, S. S., Dun, D., India. Some Studies on the Shrinkage Behaviour of Wood. *Wood Science and Technology* (1967) 1: 99-108.

Setiati, R., Wahyuningrum, D., Kasmungin, S. *Analisa Spektrum Infrared Pada Proses Sintesa Lignin Ampas Tebu Menjadi Surfaktan Lignosulfonat*, Universitas Trisakti, Seminar Nasional Cendikiawan; 2016.

Shroeder, H. A. 1992. Shrinking and Swelling Differences Between Hardwoods and Softwoods. *Wood and Fiber Science* 1: 20-25.

Silva, M. R., Pinheiro, R. V., Christoforo, A. L., Panzera, T. H., Lahr, F. A. R. 2018. Hybrid Sandwich Particleboard Made with Sugarcane, Pínus Taeda Thermally Treated and Malva Fibre from Amazon. *Materials Research* 21(1): 1-7.

Soerianegara, I., Lemmens, R. H. M. J. *Sumber Daya Nabati Asia Tenggara 5(1): Pohon penghasil kayu perdagangan yang utama*. PROSEA – Balai Pustaka. Jakarta; 2002.

Stout, A. T., Davis, A. A., Domec, J. C., Yang, C., Shi, R., King, J. S. 2014. Growth under field conditions affects lignin content and productivity in transgenic Populus trichocarpa with altered lignin biosynthesis. *Biomass and Bioenergy* 68: 228-239.

Tsoumis, G. Science and Technology of Wood. *Structure, Properties, Utilization.* New York (US): van Nostrand Reinhold; (1991).

Wibisono, H. S., Jasni, Arsyad, W. O. M. 2018. Chemical composition and natural durability of eight wood species tested under the shade. *Jurnal Penelitian Hasil Hutan* 36(1): 59-65. DOI: 10.20886/jphh.2018.36.1.59-65.

Wu, J., Fukazawa, K., Ohtani, J. 1990. Lignin Analysis in Some Tropical Hardwoods Using Ultraviolet Microscopy. *Research Bulletins of the College Experiment Forests* 47(2): 353-366.

Yang, G., Gong, Z., Luo, X. 2023. Bonding wood with uncondensed lignins as adhesives. *Nature* 621, 511–515 (2023). https://doi.org/10.1038/s41586-023-06507-5.

About the Editors

Severino G. Salmo III, PhD

Affiliation: Institute of Biology, College of Science, University of the Philippines Diliman, 1101 Quezon City, Philippines
Email: sgsalmo@up.edu.ph
Online Profile: https://mangroveecology.com
https://scholar.google.com/citations?hl=en&user=lwk4bYAAAAAJ

Dr. Salmo is an Associate Professor at the Institute of Biology, University of the Philippines Diliman. He obtained his PhD in Biological Science at The University of Queensland in Brisbane, Australia. He has a post-doc in metagenomics from the University of the Ryukyus in Okinawa, Japan. His areas of expertise are mangrove biodiversity, conservation biology, and restoration ecology. Along with his research, he works with coastal communities, NGOs, and local/national government agencies on community-based coastal management programs. He is a member of the International Blue Carbon Scientific Working Group (IBCSWG).

Dr. Salmo has several articles published in both national and international scientific journals. He is a reviewer and member of editorial boards in several journals. He has earned the NAST Environmental Science Award, two NAST Outstanding Publication Awards (in Biological Science), and was a co-awardee in two PCAARRD Elvira O. Tan Outstanding Publication Awards. He has also been awarded Professorial Chair and as UP Scientist in the University of the Philippines system. His current projects include assessment and monitoring of biodiversity, "blue carbon," and greenhouse gas emissions in conserved and restored mangroves.

Mohammad Basyuni, PhD

Affiliation: Department of Forestry; Center of Excellence for Mangrove, Universitas Sumatera Utara
Email: m.basyuni@usu.ac.id
Online Profile: https://fhut.usu.ac.id/id/dosen/mohammad-basyuni
https://scholar.google.com/citations?user=mudoMJ4AAAAJ&hl=id

Mohammad Basyuni is a Professor in the Department of Forestry, Faculty of Forestry, at the Center of Excellence for Mangrove, Universitas Sumatera Utara, Indonesia. In 2008, he graduated from the United Graduate School of Agricultural Sciences at Kagoshima University, Japan, and continued a Postdoctoral Fellow position under the support of the Japan Society for the Promotion of Science (JSPS) at the Center of Molecular Biosciences, Tropical Biosphere Research Center, University of the Ryukyus, Japan, for another two and half years. He returned to Indonesia in 2010 and continued his interest in understanding the physiological, biological, ecological, and pharmacological significance of plant lipids, including isoprenoid and long-chain polyisoprenoid in mangroves. Under this project, he collaborates with the University of the Ryukyus, Yamagata University, and Tohoku University.

One of his current research projects focuses on understanding the establishment thresholds of mangrove pioneer species in North Sumatra and their implications for restoration and climate change. The project, titled "MOnitoring Mangrove ExteNT & Services (MOMENTS): What is controlling Tipping Points?" (Joint Call Newton Fund/RCUK/LPDP/DIPI) FY2018-2023, is a collaboration with the University of Glasgow and Aberystwyth University, UK. It investigates the recovery of the macrozoobenthos community structure in relation to restoration progress and aims to build a knowledge base to support ecological restoration.

He was awarded a multilateral international joint initiative, such as the e-Asia Joint Research Program, and has worked on international-level projects with scientists from Japan (Prof. Tadashi Kajita, University of the Ryukyus) and the Philippines (Dr Venus Leopardas, Mindanao State University at Naawan). He is also a Co-Investigator of the Asia Pacific Network Global Change Research project, along with PI Dr. Alison KS Wee (University of Nottingham Malaysia), Prof. Tadashi Kajita (University of the Ryukyus), and Dr. Amy Then Yee Hui (Universiti Malaya, Malaysia). Additionally, he was a Co-Investigator for the JSPS Core to Core program from 2000-2023

(1st phase) and FY2024-2026 (2nd phase) on Global analyses of mangrove ecosystems using eDNA metabarcoding. He is currently a Scientist-Consultant of Restoration Project at Imperative, Inc.

About the Authors

Dr. Etti Sartina Siregar, SSi, Msi

Affiliation: Department of Biology Faculty of Mathematics and Sciences, Center of Excellence for Mangrove, Universitas Sumatera Utara

E-mail: etti1@usu.ac.id

Online Profile: https://scholar.google.co.id/citations?user=CLCleEsAAAAJ&hl=id

Etti Sartina Siregar is a Professor in the Department of Biology, Faculty of Mathematics and Natural Sciences, Universitas Sumatera Utara. She was born on November 21, 1972. She completed her undergraduate education at Andalas University Padang in 1996, then continued her studies at Bogor Agricultural University, where she obtained a Master of Science degree in 2004 and a doctoral degree in 2015. Since 1998, she has been a Lecturer in the Biology Department, Faculty of Mathematics and Sciences at Universitas Sumatera Utara, presenting on Plant Taxonomy (especially liverworts). She has extensive knowledge of field collecting, plant preservation and herbarium storage, given that she has a proven track record of conducting field research, maintaining herbarium collections, and contributing to scientific publications in the field of plant taxonomy and biodiversity. Today, her research focuses on liverwort phytochemicals, which are expected to have medicinal potential. She is a member of the Center of Excellence for Mangrove, Universitas Sumatera Utara and a memver of the Indonesian Society for Plant Taxonomy. She continues to research the diversity of Mangrove in Sicanang Belawan, North Sumatra.

Rahmawaty, PhD

Affiliation: Department of Forestry, Universitas Sumatera Utara
Email: rahmawaty@usu.ac.id
Online Profile: https://fhut.usu.ac.id/id/dosen/rahmawaty

Dr. Rahmawaty is a Professor in the Department of Forest Management, Faculty of Forestry and Natural Resources, and the Environmental Management Postgraduate Study Program at Universitas Sumatera Utara (USU). She completed her bachelor's degree from the Faculty of Forestry, IPB University, Bogor (1997), her postgraduate program in the Natural Resources and Environmental Management Study Program at IPB University, Bogor (2000), and earned a doctorate degree from the University of the Philippines at Los Baños (UPLB), Philippines (2009).

Currently, she is the Head of the Masters and Doctoral Study Program in Natural Resources and Environmental Management at the Postgraduate School of USU. Her field of study is related to the planning and management of forests, land resources, and the environment. She also teaches at USU's Bachelor and Master of Forestry programs, Masters and Doctorate in Agricultural Sciences, Masters in Biology, and Masters and Doctorate in Natural Resources and Environmental Management at USU's Postgraduate School.

Further, Dr. Rahmawaty is active as a researcher and expert in several government agencies and in community service activities. She actively participates in professional organizations such as the Indonesian Engineers Association (PII), Indonesian Forest Management Community, Indonesian Forestry Academic House (RAKI), Indonesian Climate Change Experts, Community Forestry Communication Forum, and Regional SEARCA Alumni Association. She is also a resource person at various seminars, training, and workshops, and she has authored several books. Additionally, she serves as an expert team member in several government and private

She has received several awards and decorations, including High Scholastic Achievement for Academic Year 2008-2009 by the Graduate School of UPLB, Los Baños, Philippines, The Honor Society of Agriculture by Gamma Sigma Delta (UPLB Chapter) at Los Baños, Philippines (2009), Outstanding Lecturer I Level, USU (2012), Life Style Achievement Award (2013) by UKWI-USU, *Satyalancana Karya Satya,* and the Pioneer Engineer Award from PII BKT HUT (2021).

Dr. Delvian

Affiliation: Department of Forestry; Center of Excellence for Mangrove, Universitas Sumatera Utara
Email: delvian@usu.ac.id
Online Profile: https://fhut.usu.ac.id/id/dosen/delvian

Dr. Delvian is a Professor in the Department of Forestry, Faculty of Forestry, at the Universitas Sumatera Utara. From 1995 – 1997, he received an URGE Scholarship from the World Bank to attend the Master of Agricultural Science Programme at Andalas University's Postgraduate Programme. With the same scholarship, he attended the Doctoral Programme in Forestry Science at Bogor Agricultural University, and graduated in 2003. Currently, he is actively conducting scientific research activities, especially in the field of plant physiology and soil microbes, which are both basic and applied studies. Today, he is active in the organization of the Indonesian Microbiology Association and the Indonesian Mycorrhiza Association. He is also active in conducting community service activities, especially related to environmental problems and issues in the forestry sector. Additionally, he often conducts technical field studies with several government and non-government institutions, including the North Sumatra Provincial Environment Office, the Indonesian Ministry of Environment and Forestry, and The Deutsche Gesellschaft für Internationale Zusammenarbeit (GIZ).

Dr. Agus Purwoko, SHut, MSi

Affiliation: Forestry Study Program, Faculty of Forestry, Universitas Sumatera Utara
Email: agus9@usu.ac.id
Online Profile: https://fhut.usu.ac.id/id/dosen/agus-purwoko

Agus Purwoko is a Professor in Natural Resource Economics at Universitas Sumatera Utara (USU). He completed his undergraduate education at the Forest Management Study Program, Faculty of Forestry, IPB University (1998), Masters at the Regional and Rural Development Planning Study Program (2005), and completed a doctoral program at the Regional Planning Doctoral Program, Postgraduate School, Universitas Sumatera Utara (2011). Dr. Purwoko has written many scientific articles and is a reviewer in various reputable national and international journals. To date, 59 Scopus-indexed articles have been recorded, and more than 300 scientific works have been recorded in Google Scholar. Most of his research and publications are related to the broad field of forestry and regional and rural development. Further, Dr. Purwoko is active in various research institutions within Universitas Sumatera Utara, especially the Center for Human-Wildlife Coexistence Studies. Dr. Purwoko collaborates extensively with researchers from various research institutions and academics in Indonesia and other countries.

Luthfi Aziz Mahmud Siregar, SP, MSc, PhD

Affiliation: Faculty of Agriculture, Universitas Sumatera Utara
Email: luthfi1@usu.ac.id
Online Profile: https://fp.usu.ac.id/id/dosen/luthfi-aziz-mahmud-siregar

Luthfi Aziz Mahmud Siregar was born in Medan on July 12th, 1973. Since 2005, he has been a Lecturer for the Faculty of Agriculture at the Universitas Sumatera Utara (USU). He earned a Bachelor of Agriculture (S1) from the Department of Agricultural Cultivation, Faculty of Agriculture, Islamic Universitas Sumatera Utara in 1996; a Master of Science in Botany (S2) in 2000, and a Doctor of Phylosophy (S3) in 2005 in Plant Cell Culture from the Institute of Post-Graduate Study, University of Sains Malaysia. His field of expertise is Plant Biotechnology: Plant Cell and Tissue Culture and In Vitro Production of Plant Secondary Metabolism. In 2006, together with Prof. Chan Lai Keng and Prof. Boey Peng Lim (Promoter and Co-Promoter), he received a Silver Medal at the Malaysia Technology Expo (MTE) for his research on the topic "Sustainable Speedy Superior Tongkat Ali." He worked as a Field Assistant at the PT UKINDO plantation (AEP Group) and as a Research Officer at the Plant Tissue & Organ Culture Laboratory, Universiti Malaysia, until he became an academic at the Faculty of Agriculture, USU. He is actively involved in several organizations, including the Indonesian Breeding Science Association (PERIPI), the Indonesian Agrotechnology/Agroecotechnology Association (PAGI), the Indonesian Horticulture Association (PERHORTI), and the Association of Indonesian Agricultural Science Doctoral Study Programs (APDISI). He is also involved in a number of research projects and community service activities, actively serving as a reviewer for national and international journals and conferences and as an expert staff member at government institutions. He has held positions as Secretary of the Plant Breeding Study Program (2009-2011) and Secretary of the Agrotechnology Study Program (2011-2027), Member of the Quality Assurance System Implementation Team at the USU Quality Management Unit Office (2018-2021), and Head of the Doctoral (S3) Agricultural Science and Masters (S2) Agrotechnology Study Program, Faculty of Agriculture, Universitas Sumatera Utara (2021-present). He has produced hundreds of articles published in a number of international and national journals and proceedings.

Mahyuddin K. M. Nasution, MIT, PhD

Affiliation: Departemen Ilmu Komputer, Fakultas Ilmu Komunikasi, Universitas Sumatra Utara
Email: mahyuddin@usu.ac.id
Online Profile: https://fasilkom-ti.usu.ac.id/id/dosen/mahyuddin

Dr. Mahyuddin K. M. Nasution, MIT, PhD, was born in Teluk Pulai Dalam, Tanjung Leidong, on December 25th, 1967. He is an alumnus of Universitas Sumatera Utara (USU), Indonesia, where he completed is undergraduate stidies in the Mathematics Study Program in 1992. He furthered his education in Information Science at Fakulti Teknologi dan Sains Maklumat (FTSM), Universiti Kebangsaan Malaysia (UKM), Bangi, Selangor, Malaysia, in 2003, and in Knowledge Technology, CAIT, FTSM, Universiti Kebangsaan Malaysia, in 2013.

He has taught various courses, including Management, Linear Programming, and Finite Element Methods in the Mathematics Study Program (Undergraduate); Databases, Programming, Machine Languages, Statistics in the Informatics Engineering Study Program (Diploma); Network Systems, Computational Linguistics in the Informatics Study Program (Magister Postgraduate); Data Modeling and Management, Business Intelligence in the Data Science and Artificial Intelligence Study Program (Magister Postgraduate); and Philosophy of Science, Ontology and Taxonomy of Information in the Computer Science Study Program (PhD Postgraduate). Prof. Nasution has been actively participating in scientific activities organized and published by IEEE, IOP, AIP, Springer, and others from 2010 to the present. He is a member of Computational Intelligence from IEEE, a member of IACSIT (International Association of Computer

Science and Information Technology) Singapore, and a member of IAENG (International Association of Engineers) Hong Kong. He served as Vice Chancellor for Research, Community Service and Cooperation at Universitas Sumatara Utara from 2015-2021.

He is also recognized as one of the world's top 2% scientists for the years 2020, 2021, and 2024 by Stanford University. To date, he has authored more than 600 scientific works (papers, articles, book chapters, and books, as well as intellectual property rights/patents), with over 150 indexed in Scopus or Web of Science (WoS) databases.

Siti Latifah, SHut, MSi, PhD, IPU

Affiliation: Department of Forestry, Universitas Sumatera Utara
Email: sitilatifah@usu.ac.id
Online Profile: https://fhut.usu.ac.id/id/dosen/siti-latifah#

Siti Latifah is a Professor in the Department of Forestry, Faculty of Forestry, Universitas Sumatera Utara, Indonesia. She completed her undergraduate study at the Faculty of Forestry, Bogor Agricultural University, IPB, in 1995. From 1997 to 2000, she received a scholarship from BPPS to continue her Masters education in the Forestry Science Program at IPB. She has been an active faculty member in the Department of Forestry at USU since 1999. In 2003, she was appointed Civil Servant in the Forestry Study Program, Faculty of Agriculture, USU. From 2006 to 2009, she took part in the Natural Resources Management Doctoral Program at the University of the Philippines Los Baños (UPLB) with scholarship funding from SEARCA SEAMEO. At USU, she has served as Dean of the Faculty of Forestry (2016-2021), Secretary of the Forest Management Study Program (2004-2006), and Chair of the Forest Management Study Program (2010-2015). During her tenure as Dean, she signed academic agreements between the USU Faculty of Forestry and the UPLB College of Forestry and Natural Resources (2018); MoU with the College of Agriculture and Life Science, Kyungpook National University (KNU), Daegu, South Korea (2017); and MoU with the Faculty of Geo-Information Science and Earth Observation (ITC), University of Twente (UT) (2019).

She is actively involved in professional organizations like the KOMHINDO Indonesian Forest Management Community, Indonesia Engineering Institute (PII), and Foretika (Deans Forum). She is also active in research and community service activities, which are funded by the USU TALENTA program, the Indonesian Collaborative Research Grant (RKI), grants from the Ministry of Education, Culture, Research and Technology, and LPDP. She serves as a reviewer for international and national journals.

Opim Salim Sitompul, MSc, PhD

Affiliation: Fakultas Ilmu Komunikasi, Universitas Sumatra Utara
Email: opim@usu.ac.id
Online Profile: https://fasilkom-ti.usu.ac.id/id/dosen/opim-salim-sitompul

Opim Salim Sitompul is a Professor at Universitas Sumatera Utara, Indonesia. With a focus on data warehouse and artificial intelligence, Sitompul has contributed to various research projects and publications in artificial intelligence and machine learning. He holds a PhD in Information Science from Universiti Kebangsaan Malaysia (UKM), Malaysia. Sitompul's current work explores AI-based approaches to evaluate students' old script writings. He is a member of the IEEE and ACM.

Ayodhia Pitaloka Pasaribu, MKed(Ped), SpA(K), PhD(CTM)

Affiliation: Department of Pediatrics, Faculty of Medicine, Universitas Sumatera Utara
Email: ayodhia@usu.ac.id
Online Profile: ORCID ID. https://orcid.org/0000-0002-3830-8073

Ayodhia Pitaloka Pasaribu was born in Banda Aceh on February 2th, 1980. She completed her Bachelor of Medicine from Universitas Sumatera Utara (1998-2004). She continued her education by pursuing a Master of Science in Pediatrics and a specialist program in pediatrics at Universitas Sumatera Utara (2005-2009), then later earned a Doctorate in Clinical Tropical Medicine from Mahidol University, Thailand (2009-2013). Since 2005, she has been a Lecturer at the Faculty of Medicine of Universitas Sumatera Utara. Over the last 20 years, she has served as Vice Chair of the Malaria National Expert Committee, Chair of the North Sumatra Diptheriae Expert Committee, member of the Indonesian Technical Advisory Group on

Immunization, member of the Indonesia College of Pediatrics, and many more. Currently, she is Head of the Pediatrics Specialist Program at the Faculty of Medicine of Universitas Sumatera Utara (2021-2026).

She is an active researcher with over 40 publications in renowned journals; she has conducted research in collaboration with many institutions, including Oxford University, Menzies School of Health Research, Lyon University, École Pratique des Hautes Études, and more. Further, she is an active reviewer for several journals, such as *Lancet Infectious Diseases, PLos One, Malaria Journal,* and is an editor for *BMC Infectious Diseases*. Ayodhia continues to excel in her primary roles as Professor of Pediatrics and Senior Lecturer in Infectious Diseases and Tropical Medicine.

Dr. Dewi Masyithah Darlan, DAP&E, MPH, SpParK

Affiliation: Department of Parasitology, Faculty of Medicine, Universitas Sumatera Utara
Email: dewi2@usu.ac.id
Online Profile: https://fk.usu.ac.id/id/dosen/dewi-masyithah-darlan#

Dewi Masyithah Darlan is a Professor in the Department Parasitology, Faculty of Medicine, Universitas Sumatera Utara. She was born in Medan on July 30th, 1974. She completed her undergraduate study in Medicine from USU Medan (1992 to 1998). Following this, she worked as a Lecturer at Universitas Sumatera Utara until 2001. She continued her education by pursuing a post graduate diploma at the Institute of Medical Research (May – October 2003) and earning a Master of Public Health from Southern Cross University (2004-2005) as well as a Master of Clinical Parasitology from Universitas Indonesia (2011-2014).

From 2016 to 2020, Dewi obtained her Doctor of Medicine from Universitas Sumatera Utara. She has held various academic leadership roles, including Head of the Medical Education Unit and Head of the Undergraduate Medicine Study Program. She is actively involved in professional organizations like IAMCP (Clinical Parasitology Indonesia), MSPTM (Malaysian Parasitology) and IDI (Indonesia Medical Association).

Fahmi, ST, MSc, PhD, IPM, ASEAN Eng

Affiliation: Department of Electrical Engineering, Faculty of Engineering, Universitas Sumatera Utara
Email: fahmimn@usu.ac.id
Online Profile: https://ft.usu.ac.id/id/dosen/fahmi

Dr. Fahmi, born in Medan on December 9th, 1979, completed his undergraduate study in Electrical Engineering from ITB Bandung (1998 to 2002). He continued his education by pursuing a master's degree in Sensor System Technology at FH Karlsruhe, Germany (2003-2005). Following this, he worked in the medical equipment field as a product manager at a private company, then transitioned into an academic role as a lecturer at Universitas Sumatera Utara (2007). During this time, he actively contributed to the development of RSUSU Medan Hospital (2007-2010). From 2010 to 2015, Dr. Fahmi obtained his Ph.D. in Biomedical Engineering from Universiteit van Amsterdam, the Netherlands. He has held various academic leadership roles, including Head of COMLABS Laboratory, Head of the Electrical Engineering Department, Active Senator of the Academic Senate (SA) at USU, and a member of the Board of Trustees (MWA) at USU. He is actively involved in professional organizations like IEEE (Senior Member), PII (Indonesia Engineering Institute), and FDTI (Deans Forum). He is also involved in various projects as a professional engineer, actively becoming a reviewer and assessor for ministry programs and accreditation boards (national and international) as well as journals and reputable conferences, and he has become an expert staff member for government institutions. Currently serving as Dean of the Faculty of Engineering at USU, Dr. Fahmi continues to excel in his primary roles as a senior lecturer and researcher in digital technology, specifically in electrical engineering and biomedical engineering.

Dr. Abdul Rauf, MP

Affiliation: Faculty of Agriculture, Universitas Sumatera Utara
Email: a.raufismail@gmail.com
Online Profile: https://fp.usu.ac.id/id/dosen/abdul-rauf

Abdul Rauf was born in Binjai on September 17th, 1959. He is a Professor of Soil Conservation and Watershed Management at the Faculty of Agriculture, Universitas Sumatera Utara (USU). In 1985, he earned a bachelor's degree from the Soil Science Study Program, Faculty of Agriculture, Universitas Sumatera Utara (USU). In 1994, he earned a master's degree in soil science at UGM Yogyakarta. He then continued his education and, in 2004, earned a doctorate degree in the same field at IPB Bogor. Since 1986, he has studied the field of soil science, specifically soil and water conservation and watershed management; he became a member of the teaching staff at several universities in Medan in this field. In 2004, he completed his service as a member of the teaching staff in the Postgraduate Masters and Doctoral Program in Agricultural Science, Faculty of Agriculture, USU.

In addition, Dr. Rauf has written several textbooks: *Basics of Watershed Management* (USU Press); *Agroforestry Systems, Sustainable Land Empowerment Efforts* (USU Press, 2011) and its revised edition (USU Press, 2017); *USU Arboretum Profile 2006-2008* (USU Press, 2009); *Watershed Management, An Integrated Management Plan for the Asahan Toba Watershed* (USU Press); *Watershed Management, An Integrated Management Plan for the Batang Gadis Watershed* (CI-Jakarta); *Land Utilization Technology Based on Watershed Management* (USU Press, 2016); *Acid Sulfate Soil and Its Biological Management* (USU Press); *Sustainable Oil Palm Cultivation* (USU Press, 2018); *Optimization of Agricultural Land Using Biomass Agents* (USU Press, 2019); *North Sumatra Wampu Watershed Expedition* (USU Press, 2019); *Exploring the Singkil Watershed from Upstream to Downstream* (USU Press, 2019); *Rice Cultivation in Saline Land* (UMSU Press, 2020); *Biophysical Science and Agriculture in the Perspective of the Qur'an* (USU Press, 2020); *Black Rice* (UMSU Press, 2023); and *Conservation of the Lake Toba Catchment Area* (CV. Mega Press Nusantara, 2024). Furthermor, Dr. Rauf is involved in several professional organizations, such as the Central Management of the Indonesian Soil and Water Conservation Society (MKTI), the Management of the North Sumatra Regional Commissariat of the Indonesian Soil Science Association (HITI), the Indonesian Peat Association (HGI), and the North Sumatra Wampu Watershed Forum.

Dr. R. Hamdani Harahap MSi

Affiliation: Department of Anthropology, Faculty of Social and Political Sciences (FISIP), University of Indonesia
Email: r.hamdani@usu.ac.id
Online Profile: https://fisip.usu.ac.id/id/dosen/r.-hamdani-harahap

Dr. Harahap was born in Pematang Siantar on February 27th, 1964. He obtained his bachelor's degree from the Department of Anthropology, FISIP, USU in 1988, followed by a master's degree in Natural Resources and Environmental Management (PSL) from IPB Bogor in 1994. He completed his doctoral studies in the Natural Resources and Environmental Management Program at USU, Medan, in 2009. From 1995 to 2000, he served as the Chair of the Social Anthropology Department at USU's FISIP. Then, he held the position of Assistant Dean III for Student Affairs at USU's FISIP from 1999 to 2005. He later worked as a Training Specialist at the United States Agency for International Development Local Governance Support Program, NSRO, from 2005 to 2009. Furthermore, he participated in Leadership Training at the University of Arkansas, USA from October to December 2009 through the Ford Foundation. From 2010 to 2021, he served as the Secretary of USU's FISIP Master's Study Program and was appointed as a Professor of Human Ecology at USU's FISIP in 2014. His research has primarily focused on ecology and culture, particularly in coastal ecosystem areas. Since 2016, he has been conducting research related to Lake Toba's ecosystem. From May to July 2024, he was on sabbatical leave from the University of South Dakota, USA to research coffee in North Sumatra and Aceh Province, as well as coffee shops in Medan.

Dr. Suhaidi, SH, MH

Affiliation: Faculty of Law, Universitas Sumatera Utara
Email: suhaidi@usu.ac.id
Online Profile: https://fh.usu.ac.id/id/dosen/suhaidi
https://scholar.google.co.id/citations?user=_LQ_CEUAAAAJ&hl=id

Dr. Suhaidi is a Professor at the Faculty of Law, Universitas Sumatera Utara, Indonesia. In 2002, he earned his PhD from the Faculty of Law, Universitas Sumatera Utara by defending his dissertation entitled 'Protection of the Marine Environment from Ship-Source Pollution: Consequences of the Application of International Navigation Rights through Indonesian Waters.' Currently, he is a lectuer and a member of the Board of Professors at Universitas Sumatera Utara, a member of the Indonesian Society of International Law Lecturers, an expert at the Environmental Agency (BLH), Environmental Department (DLH) of North Sumatra Province (2012 - present); and an Expert Council Member of the Indonesian Environmental Scholars Association (2021 - 2026). Furthermore, he is a recognized expert in environmental and international law. His most recent research focuses on the role of the Indonesian government in controlling illegal, unreported and unregulated (IUU) fishing in the Malacca Strait after the implementation of the IUU fishing policy in 2021 (Study: Marine Environment in North Sumatra Province). His latest publications are *State Responsibility for Forest and Land Fires in Indonesia*, published by CV. Merdeka Kreasi Group (December, 2023) and *Corporate Legal Responsibility Mineral and Coal Mining Environmental Rehabilitation After Revocation of Mining License as An Effort to Sustainable Development*, published by CV. Eureka Media Aksara (2024).

Dr. Sunarmi, SH, M. Hum

Affiliation: Faculty of Law, Universitas Sumatera Utara
Email: sunarmi@ac.id
Online Profile: https://fh.usu.ac.id/en/lecturer/sunarmi

Dr. Sunarmi is a Professor in the Department of Economic Law, Faculty of Law, Universitas Sumatera Utara. Since 1989, she has taught undergraduate, master's and doctorate programs. She is an expert in the field of Business Law. In 2005, she earned her PhD from the Faculty of Law, USU. From 2011-2014 she served as Dean of the Faculty of Law, University of Riau Pekanbaru and has since taught at the University of Riau.

Additionally, she serves as an Honorary Council Member of North Sumatra in the Indonesian Advocates Association. She actively writes and conducts research in the fields of environment and business, and she has published several journals and books in those fields. Her most renowned book is on the topic of business law, particularly bankruptcy and money laundering, and is entitled *Principles of Customer Due Diligence in Money Laundering Crimes in Indonesia* (2023). Moreover, her most recent publication is a book entitled *Trademark Registration Book (An Introduction)* (February, 2024).

Himsar Ambarita, PhD

Affiliation: Sustainable Energy Research Centre, Universitas Sumatera Utara
Email: himsar@usu.ac.id

Dr. Himsar Ambarita, a distinguished professor at Universitas Sumatera Utara (USU), is at the forefront of sustainable engineering. With initial disinterest in farming, he found his passion at USU's Faculty of Mechanical Engineering, where he thrived academically and socially. His career in academia ignited a deep passion for teaching and research, leading him to become a lecturer and civil servant at USU. Himsar's quest for knowledge took him to Japan, where he earned a doctoral scholarship to study sustainable energy and climate change. His research focuses on renewable energy systems, particularly solar and bioenergy technologies.

Professor Ambarita has made significant contributions to the field through numerous research papers published in high-impact journals and active participation in international conferences. His innovative work addresses energy efficiency and sustainability challenges, bringing practical applications to life. At USU, he plays multiple roles, advancing engineering education and fostering a culture of innovation and excellence. Professor Ambarita's dedication to mentoring young engineers and engaging in research projects funded by national and international organizations promotes green technology and sustainable development. His efforts to integrate sustainable practices into engineering solutions continue to shape a greener future for Indonesia and beyond.

Ilmi Abdullah, MSc, PhD

Affiliation: Faculty of Engineering, Universitas Sumatera Utara
Email: ilmiabdullah@itm.ac.id
Online Profile: https://ft.usu.ac.id/id/dosen/ilmi

Dr. Abdullah was born on October 7th, 1959. Currently, he serves as a Lecturer in Mechanical Engineering at the Faculty of Engineering, Universitas Sumatera Utara. He completed his doctorate degree from the Faculty of Engineering at Universiti Kebangsaan, Malaysia, in 2003.

Zulkifli Nasution, MSc, PhD

Affiliation: Faculty of Agriculture, Universitas Sumatera Utara
Email: zulnasution@usu.ac.id
Online Profile: https://fp.usu.ac.id/id/dosen/zulkifli-nasution

Zulkifli Nasution was born in Kisaran on August 15th, 1959. He is an alumnus of the Universitas Sumatera Utara's Faculty of Agriculture (1984), Ghent University's Department of Geology (1989), and Universiti Sains Malaysia (2004). Dr. Nasution's passion for learning has led him to attend several courses, including Photogrammetry and Remote Sensing in the United Nations Development Programme (UNDP) at Bandung Institute of Technology in 1987, Land for Development at ITC and and Landbow Watenschapen (Agricultural Sciences) at Ghent University in 1988, Wetland Management at the Department of Resource Development of Western Australia in 1994, and Ziel Orienterte Projekt Planung (Goal-Oriented Project Planning) Training at GTZ in 1998.

In 2004, Dr. Nasution conducted environmental impact analyses on coastal zone habitat issues and case studies at the Muca Head Marine and Coastal Research Station. In 2007, he attended the University Administration and Management Program at Auburn University, Alabama, to further hone his skills in leadership and management. Similarly, in 2013, he attended the Leadership Development Program at Glasgow University, Scotland.

Dr. Marheni, MP

Affiliation: Faculty of Agriculture, Universitas Sumatera Utara
Email: marheni@usu.ac.id
Online Profile: https://fp.usu.ac.id/id/dosen/marheni

Dr. Ir. Marheni was born in Medan on July 24th, 1965. Since 1989, she has been a lecturer at the Faculty of Agriculture, Universitas Sumatera Utara

(USU). Majoring in plant protection, she earned a Bachelor of Agriculture from the Faculty of Agriculture, Universitas Sumatera Utara in 1988. In 1993, she earned a master's degree from the Plant Pest Science Study Program, Faculty of Agriculture, University of Gadjah Mada, Yogyakarta. Further, in 2012, she earned a doctorage degree in the field of plant pests from the Doctoral Program of Agricultural Sciences, University of Gadjah Mada, Yogyakarta. Her area of expertise is in entomology. In 2021, she conducted entomological research entitled "Environmental Safety Testing of Staked Event PRG Corn in the Limited Test Field (LUT)". The research was carried out by PT Syngenta Seed Indonesia in collaboration with Universitas Sumatera Utara. She received the "Best Paper Award" at the International Plantation Conference (IPC) in 2021.

T. Sabrina, MAgrSc, PhD

Affiliation: Faculty of Agriculture, Universitas Sumatera Utara
E-mail: t.sabrina@usu.ac.id
Online Profile: https://fp.usu.ac.id/id/dosen/t-sabrina

Dr. T. Sabrina is a professor of Agrotechnology at Universitas Sumatera Utara who serves as Director of the Postgraduate School and Chief of Sumatera Utara' Soil Science Society. As an expert in soil science, she has made significant contributions to the development of agricultural technology, particularly in soil science and land resource management. Professor Sabrina earned her PhD in Land Resource Management from Universiti Putra Malaysia (UPM).

She has dedicated her academic career to advancing knowledge in agrotechnology, focusing on understanding and managing soil science. Her commitment to education, research, and community service reflects her dedication to improving agricultural sustainability in North Sumatra and beyond. As a professor, she has played a pivotal role in teaching, mentoring students, and conducting impactful research aimed at promoting better agricultural practices and sustainable land use. With her international academic training and extensive experience in Indonesia, Professor Sabrina is recognized as a leading figure in agricultural sciences, making valuable contributions to improving land resource management techniques in the region. Additionally, she actively engages in community service, empowering farmers through training and development projects, reinforcing her influence in agriculture and commitment to enhancing farmers' livelihoods in Indonesia and beyond.

Hotnida Sinaga, PhD

Affiliation: Faculty of Agriculture, Universitas Sumatera Utara
Email: hotnida@usu.ac.id
Online Profile: https://fp.usu.ac.id/id/dosen/hotnida-sinaga

Hotnida Sinaga is a Professor of Food Technology at Universitas Sumatera Utara, specializing in food processing, since 2023. Her current position is Head of the Food Science Study Program, where she ensures a high-quality learning experience for master's students by attending to academic issues and assisting students with their research projects. Hotnida earned an engineering degree from Universitas Sumatera Utara and was appointed as a teaching staff member at the Faculty of Agriculture. She received a Master of Philosophy (MPhil) degree from the School of Land, Agriculture and Food Sciences, The University of Queensland, Australia. She later received a Doctor of Phylosophy (PhD) degree from the same university. Hotnida is active in conducting research, community service and writing scientific articles in the fields of food science and technology, and she is also a resource person at various national and international activities. Her research focuses on developing novel processing and preservation strategies that enhance the safety, quality, and nutritional value of food products. Hotnida has produced multiple patents and has published many scientific publications in accredited national journals, international proceedings indexed by Scopus and reputable journals indexed by Scopus. Her work has also been peer-reviewed in various international journals indexed by Scopus and WoS.

Dr. Darma Bakti

Affiliation: Faculty of Agriculture, Universitas Sumatera Utara
Email: darma@usu.ac.id
Online Profile: https://fp.usu.ac.id/id/dosen/darma-bakti

Darma Bakti earned his bachelor's degree from the Faculty of Agriculture and his doctorate degree in agriculture entomology from the

Agricultural Sciences Program at Gajah Mada University. During his career at the Faculty of Agriculture, he researched food security, especially food crops in North Sumatra. For 14 years, he was involved in training urban insects via pest control. For 10 years, he was Head of Farmer Extension Services in North Sumatra.

He has spoken at several forums regarding food security and providing guidance and services to farmer extensions in North Sumatra. He has written several books and papers for students and entomological associations. Furthermore, he is a member of national entomological societies as well as economical and agricultural societies in Indonesia.

Dr. Rosmayati

Affiliation: Faculty of Agriculture, Universitas Sumatera Utara
Email: rosmayati@usu.ac.id
Online Profile: https://fp.usu.ac.id/id/dosen/rosmayati

Dr. Rosmayati completed her undergraduate studies at the Faculty of Agriculture, Universitas Sumatera Utara, in 1983 and earned her PhD degree from IPB University in 2009. Her research interests include food crop breeding practices and agronomy. She has spoken and participated in several seminars and conferences. Her postdoctoral studies include tissue culture (Gajah Mada University) and agroclimatology cooperation (IPB University and SEAMEO). Currently, she is a Head Professor at Universitas Sumatera Utara.

Dr. Noverita Sprinse Vinolina, MP

Affiliation: Faculty of Agriculture, Universitas Sumatera Utara
Email: noverita@usu.ac.id
Online Profile: https://fp.usu.ac.id/id/dosen/noverita-sprinse-vinolina
https://scholar.google.co.id/citations?user=FjePIugAAAAJ&hl=id

Noverita Sprinse Vinolina is a Professor at the Department of Agrotechnology, Faculty of Agriculture, Universitas Sumatera Utara, Indonesia. She completed her undergraduate studies in the Faculty of Agriculture at Universitas Sumatera Utara (USU) in 1992. She obtained her

Master of Agriculture degree from the Agronomy Study Program at the Postgraduate School of Universitas Sumatera Utara in 1997, and her Doctoral Degree from the Faculty of Agriculture, at the same university in 2014. All levels of her education were funded by the Directorate General of Higher Education of the Republic of Indonesia.

In 2011, she had the opportunity to participate in the Sandwich-like Program funded by the Directorate General of Higher Education of the Republic of Indonesia at the Department of Biology/Molecular Biology and Pharmacy at Auburn University, USA. There, she conducted research on the growth, biomass production, and secondary metabolite content of centelloside with Professor Narendra K. Singh and Professor Jay Ramapuram.

Dr. Vinolina actively engages in educational, research, and community service activities. She has received several research grants and has numerous publications focused on production enhancement technologies in food crops. Her research is related to agronomic practices and agrotechnology to increase growth and the production of plants, including medicinal plants in Indonesia such as pegagan (*Centella asiatica*) and its secondary metabolic content. Her work includes the cultivation of several accessions of pegagan from North Sumatra and its centelloside content. She has also conducted research on the life cycle assessment of shallot farming in Food Estate Hutajulu, North Sumatra, Indonesia, and analyzed the energy used in shallot farming in the same area, funded by the Ministry of Education, Culture, Research and Technology.

She is also active as a resource person at scientific meetings and workshops, a presenter at international and national seminars, and an author of textbooks (ISBN). Additionally, she is actively involved in professional organizations such as PII (Indonesia Engineering Institute) and PERAGI (Indonesian Agronomy Association).

Dr. Tavi Supriana, MS

Affiliation: Department of Agribusiness; Faculty of Agriculture, Universitas Sumatera Utara
Email: tavi@usu.ac.id
Online Profile: https://fp.usu.ac.id/id/dosen/tavi-supriana
https://scholar.google.com/citations?user=jbUZUFcAAAAJ&hl=en

Dr. Suprian was born in Pematang Siantar on November 2nd, 1964. She obtained a Bachelor of Agriculture degree from Universitas Sumatera Utara (USU) in 1988, a master's degree from the USU Medan Postgraduate Program 1995, and a Doctorate degree from the Agricultural Economics Study Program of the Graduate School at Bogor Agricultural University (IPB) in 2004. Dr. Supriana is currently a lecturer in the Agribusiness Study Program.

She has held several positions, including Chairperson of the Agribusiness Study Program (2006-2008), Chairperson of the Agribusiness Master Study Program (2009-2016), Vice Dean III for Research, Service, and Cooperation of the Faculty of Agriculture at USU (2016-2021), and Dean of the Faculty of Agriculture at USU (2021-2026). Additionally, Dr. Supriana is active in professional organizations such as the Indonesian Agricultural Economics Association (PERHEPI), the Indonesian Economists Association (ISEI), the AAI (Indonesian Agribusiness Association), and the International Society for Southeast Asian Agricultural Sciences (ISSAAS). Furthermore, she has been Treasurer of FKPTPI (2023-2025) and Secretary of BKS PTN Barat (2023-2025). Dr. Supriana is regularly invited to speak at national forums and serves as a reviewer of several national and international published journals. Additionally, she has various publications—both internationally reputable and nationally accredited—and has published five reference books: *Nonparametric Statistics: Applications in Agricultural Socioeconomics*; *Socioeconomic Research Methods*; *Rice Consumers: Preferences and Willingness to Pay*; *Indonesian Business Cycle Analysis*; and *Macroeconomics*. She has received many research grants and publications focusing on agricultural economics, particularly in the field of food and tubers.

Saharman Gea, SSi, MSi, PhD

Affiliation: Department of Chemistry, Faculty of Mathematics and Natural Sciences, Universitas Sumatera Utara
Email: s.gea@usu.ac.id

Saharman Gea is a Professor of Physical Chemistry of Materials at the Faculty of Mathematics and Natural Sciences, Universitas Sumatera Utara. He was born in Lahewa on November 10th, 1968. He earned his bachelor's degree in chemistry from Universitas Sumatera Utara (USU) in 1997 and completed earned master's degree in chemistry at the same university in 2000. In 2011, he obtained his PhD from the School of Engineering and Materials Science at Queen Mary University of London. From 2011 to 2012, he was a visiting professor at Toyohashi University, Japan. Dr. Gea has been an active faculty member in the Chemistry Department at USU since 1999. His academic contributions at USU include serving as Vice Dean III of the Faculty of Mathematics and Natural Sciences (2016–2021), Head of the Center for Cellulose Research, and an active senator in the USU Academic Senate. Beyond his roles at USU, he has also served as an expert advisor to the Regent of North Nias, North Sumatra (2013–2016) and as an advisor to the Rector of Universitas Prima Indonesia. As a researcher, Saharman Gea has been involved in various projects regarding biomaterials, renewable energy, wound dressings, and catalysts. He has over 106 publications in reputable international journals, with a Scopus h-index of 18, and he has authored several books. Furthermore, he received the Best Lecturer-Writer Award from USU, and his expertise has led him to be a speaker at numerous international and national conferences as well as a reviewer for international journals and scholarship programs under Indonesia's Ministry of Education.

Dr. Tamrin, MSc

Affiliation: Faculty of Mathematics and Natural Sciences, Universitas Sumatra Utara
Email: thamrin@usu.ac.id
Online Profile: https://fmipa.usu.ac.id/id/dosen/tamrin

Dr. Tamrin was born in Central Aceh on July 4th, 1960. He is a lecturer in Chemistry at the University of North Sumatra. He earned his PhD in chemistry from Universitas Sumatra Utara in 2007.

Andriayani, SPd, MSi

Affiliation: Faculty of Mathematics and Natural Sciences, Universitas Sumatra Utara
Email: andriayani@gmail.com
Online Profile: https://fmipa.usu.ac.id/id/dosen/andriayani

Dr. Andriayani was born in Takengon on March 5th, 1969. He is a lecturer in Chemistry at Universitas Sumatra Utara. He earned his PhD in chemistry from Universitas Sumatra Utara in 2012.

Dr. Elisa Julianti

Affiliation: Department of Food Science, Center of Tuber and Root Crops Study, Universitas Sumatera Utara.
Email: elisa1@usu.ac.id
Online Profile: https://fp.usu.ac.id/id/dosen/elisa-julianti
https://scholar.google.com/citations?user=VzFL9sAAAAAJ&hl=id

Elisa Juliantti is a professor in Food Technology at the Faculty of Agriculture, Universitas Sumatera Utara. She was born in Medan on July 16th, 1967. She earned her bachelor's degree in Agricultural Product Technology from Universitas Sumatera Utara (USU) in 1990, completed her master's degree in Postharvest Technology at IPB University in 1997, and obtained her PhD in Food Science from IPB University in 2002. In 2017, she participated in a Visiting Lecturer program at Universiti Sains Malaysia.

Elisa Juliantti has been a faculty member in the Department of Agricultural Product Technology at Universitas Sumatera Utara since 1991 and is currently based in the Master's Program of Food Science at the Faculty of Agriculture, USU. Her contributions at USU include serving as Head of the Master's Program in Food Science from 2012-2016, Head of the Undergraduate Food Technology Program from 2016-2021, and Vice Dean III of the Faculty of Agriculture from 2021-2026. She also serves as Head of the Center for Tuber and Root Crops Studies at USU.

Her research interests focus on food processing technology, particularly postharvest technology, drying technology, processing of flour and starch from tuber crops, and functional foods. She has actively published her work in various scientific journals, both national and international, with 77 publications in reputable international journals and over 100 publications in national journals. She has also authored several books.

She is an active reviewer for numerous international scientific journals, reviewer for research projects at USU's Research Institute and also reviewer for community service program at Directorate of Research, Technology, and Community Service, Ministry of Education, Culture, Research and Technology Republic of Indonesia since 2011. Elisa Juliantti is also an active member in several professional organizations, including the Indonesian Association of Food Technologists, the Indonesian Functional Food and Nutraceutical Enthusiasts Association, and the Indonesian Researcher Association.

Dr. Diana Sofia Hanafiah

Affiliation: Faculty of Agriculture, Universitas Sumatera Utara
Email: diana.hanafiah@usu.ac.id
Online Profile: https://fp.usu.ac.id/id/dosen/diana-sofia-hanafiah

Diana Sofia Hanafiah was born in Medan on August 30th, 1974. She has been a lecturer at the Faculty of Agriculture, Universitas Sumatera Utara (USU) since 1999. She completed her Bachelor of Agriculture at the Department of Plant Breeding, Faculty of Agriculture, USU, in 1997, Master of Agriculture at the USU Postgraduate Program in 2001, and Doctor of Agriculture at the Plant Breeding and Biotechnology Study Program, IPB Graduate School, in 2012. Her areas of expertise are Plant Breeding and Mutation. Dr. Hanafiah has been awarded the Twenty-Fourth Best Paper Writer in the Accredited National Periodical Category at the National Seminar of the Indonesian Breeding Science Association (PERIPI) in Bogor in 2012. She has also gained intellectual property in the form of seven patents and two copyrights. Additionally, Dr. Hanafiah is a reviewer for several national and international journals. She has written several books, including *Acid Sulfate Soil and Its Biological Management*, *Environmentally Friendly Development of Indonesia in the Era of Globalization: Multidisciplinary Studies*, and *Drought Stress Breeding in Soybean*. Moreover, she is part of the variety verification team in the context of releasing plants as varieties, appointed by the North Sumatra Horticulture Plant Department in 2024.

Dr. Yaya Hasanah, MSi

Affiliation: Faculty of Agriculture, Universitas Sumatera Utara
Email: yaya@usu.ac.id
Online Profile: https://fp.usu.ac.id/id/dosen/yaya-hasanah

Yaya Hasanah is a Professor in Agronomy at the Faculty of Agriculture, Universitas Sumatera Utara (USU). She was born in Ciamis on January 10th, 1969. She graduated from the Undergraduate Agronomy Study Program (S1), Faculty of Agriculture, Institut Pertanian Bogor in 1992. She earned her Master of Science (S2) in 2003 from the Postgraduate Program at Universitas Sumatera Utara and completed her Doctoral degree (S3) in Agricultural Sciences at the Faculty of Agriculture, Universitas Sumatera Utara in 2014.

In 2010, she participated in the Sandwich like-Program at Auburn University, Alabama, USA. She plays an active role in teaching in the Bachelor's and Master's study programs in Agrotechnology and the Doctoral Programs in Agricultural Sciences at the Faculty of Agriculture, USU. She is actively involved in research, community service, scientific publications, and as a speaker and participant in national seminars, international conferences, and training/workshops.

Her scientific works and research results are published in various international journals indexed by Scopus/WoS, proceedings indexed by Scopus, and accredited national journals. She is also a peer-reviewer in various international journals indexed by Scopus and WoS. In 2017, Dr. Hasanah was a speaker at "Tropen Tag 2017: International Research on Food Security, Natural Resource Management and Rural Development" at the University of Bonn, Germany.

Furthermore, Dr. Hasanah has written several books, including *Cultivation of Medicinal Plants and Spices* and *Post-Harvest Cultivation and Technology of Ginger Harvest*. She is a member of many professional organizations, including the Indonesian Agroecotechnology/Agrotechnology Association (PAGI), the Indonesian Horticultural Association (PERHORTI), the Indonesian Agricultural Meteorology Association (PERHIMPI) and the Indonesian Agronomy Association (PERAGI).

Irvan, MSi, PhD

Affiliation: Department of Chemical Engineering, Sustainable Energy and Biomaterial Center of Excellence, Universitas Sumatera Utara
Email: irvan@usu.ac.id
Online Profile: https://ft.usu.ac.id/id/dosen/irvan

Professor Irvan was born in Tebing Tinggi on August 20th, 1968. He is currently a professor in the Chemical Engineering Department, at the Sustainable Energy and Biomaterial Center of Excellence, Universitas Sumatera Utara. He has held several academic positions, including Head of Physical Chemistry Laboratory, Head of Chemical Engineering Department, Vice Dean for Research and Collaboration Affair, Academic Senator of USU, and member of the Board of Trustees at USU. In addition to his academic positions, he is involved in various projects, particularly focusing on the environment and energy.

In 2006, he graduated from the Environmental and Life Engineering Doctoral Course at Toyohashi University of Technology, Japan under the Monbukagaskusho Scholarship. His dissertation was entitled "Application of Supercritical CO_2 Extraction Technology to the Determination of Bacterial Quinones in Environmental Sample." He has since collaborated with various institutions, including Toyohashi University of Technology, Metawater Co. Ltd., Technische Universiteit Eindhoven, Boras University, Kita Kyushu University, and TU Berlin, to conduct joint research in Biogas Generation from Plam Oil Mill Effluent (POME).

Seri Maulina MSi, PhD

Affiliation: Research Cluster for Life Cycle Assessment, Universitas Sumatera Utara
Email: serimaulina@usu.ac.id
Online Profile: https://ft.usu.ac.id/id/dosen/seri-maulina

Seri Maulina was born in Medan on January 4th, 1961. She is currently a professor and lecturer in the Department of Chemical Engineering at the Faculty of Engineering, Universitas Sumatera Utara. She earned her bachelor's degree in chemical engineering from Universitas Sumatera Utara in 1986, her master's degree from Bandung Institute of Technology in 1994, and her PhD from the University of Malaya in 2014. She has been a lecturer in the Chemical Engineering Department since 1988. From 2021 to 2024, she served as a member of the Executive Board of the Independent Accreditation Institute for Engineering Study Programs. Since 2023, she has led the Research Cluster for Life Cycle Assessment at Universitas Sumatera Utara. Previously, she has served as Dean of the Faculty of Engineering

(2016 to 2021), Secretary of the Chemical Engineering Department (1996 to 2000), and Head of the Organic Laboratory (1995 to 1997).

Outside USU, she has held several significant positions. She served as an expert staff member and team leader for the Environmental Section of the Medan Municipality from 1997 to 2005, and as team leader for testing chimney emissions, ambient air, and liquid waste at the Palm Oil Mill of Sumatera Plantation, Tanjung Kasau from 2005 to 2010. Additionally, she was the Environmental Audit Coordinator for PT (Persero) Indonesia Port 1, overseeing ports such as Malahayati Port, Kuala Langsa Port, Selat Panjang Port, Bengkalis Port, Kuala Tanjung Port, Bagan Asahan Port, Gunung Sitoli Port, Sibolga Port, Tembilahan Port, and Belawan Port from 2006 to 2010. Currently, she is conducting research related to pyrolysis, advanced processes, and life cycle assessment in the palm oil industry.

Dr. Renita Manurung, MT

Affiliation: Department of Chemical Engineering, Faculty of Engineering, Universitas Sumatera
Email: renitachem@yahoo.com
Online Profile: https://ft.usu.ac.id/id/dosen/renita-manurung

Dr. Renita Manurung was born in Pematangsiantar on December 14th, 1968. Currently, she is a lecturer in the Department of Chemical Engineering at the Faculty of Engineering, Universitas Sumatera Utara (USU). Dr. Renita has a strong educational background in chemical engineering, renewable energy, and oleochemistry. She completed her undergraduate degree (S1) in 1992 in Chemical Engineering, pursued a master's degree (S2) in 2005, and earned her doctoral degree (S3) in the same field in 2017 at Universitas Sumatera Utara. This academic journey reflects her deep commitment to knowledge and research in chemical engineering, particularly in renewable energy and biodiesel.

As a lecturer, Professor Renita is not only involved in teaching but also in research, especially in developing sustainable and environmentally friendly chemical technologies. Her research areas include renewable energy utilization, waste treatment, and the development of efficient and sustainable chemical processes. One of her most significant studies focuses on "Eutectic Solvent Technology and Integrated Palm Biodiesel Production Technology with Palm Micronutrients (Carotene and Tocopherol-Tocotrienol) as an

Alternative Solution to Promote the Growth of the National Biodiesel Industry."

Furthermore, Dr. Renita has written and published many scientific papers in international journals as well as patent publications, and she has participated in various collaborative research projects both domestically and internationally. With her extensive academic experience, she is frequently invited to speak at seminars, conferences, and workshops. In addition to her duties as a lecturer in fulfilling the Tridharma of Higher Education, Professor Renita has also served as Head of the Chemical Engineering Study Program. She currently serves as Vice Dean 1 (2021-2026). As an educator and researcher, Professor Renita is well-known for her strong dedication to developing younger generations.

Tulus, VorDiplMath, MSi, PhD

Affiliation: Fakultas Matematika dan Ilmu Pengetahuan Alam, Universitas Sumatera Utara
Email: tulus@usu.ac.id
Online Profile: https://www.researchgate.net/profile/Tulus-Tulus

Tulus studied in the Department of Mathematics Universitas Sumatera Utara 1987 and earned his Master's degree in 1996 from Institut Teknologi Bandung. He received his PhD degree in 2007 from Universiti Kebangsaan Malaysia under the supervision of the late Prof. Dr. Ahmad Kamal Ariffin. His research interests include Applied Mathematics, Computational Fluids Dynamics, Finite Element Method, Numerical Analysis, Optimization, Complex Analysis, Simulation, and Fatigue Analysis. He has written 94 articles published in peer-reviewed journals and proceedings, with a Scopus h-index of 10 and a Google Scholar h-index of 14. Since 2007, he has supervised and co-supervised PhD students.

Rosdanelli Hasibuan, MT, Ph.D

Affiliation: Department of Chemical Engineering, Faculty of Engineering, Universitas Sumatera Utara
Email: rosdanelli@usu.ac.id
Online Profile: 19933430300

Rosdanelli Hasibuan was born in Pangkalan Brandan on August 8th 1968. She completed her undergraduate studies in Chemical Engineering from Universitas Sumatera Utara, Indonesia (1993). She continued her education by pursuing a master's in Chemical Engineering at Institut Teknologi Bandung, Indonesia (1998). Following this, she obtained her PhD from the Department of Chemical and Process Engineering at Universiti Kebangsaan Malaysia, Malaysia (2006). Then, she received a Professional Engineering Certificate from the Engineer Professional Study Program (PPI) at Universitas Sumatera Utara (2020).

She has held various academic leadership roles, including Head of Chemical Industrial Process Laboratory and active senator of the Academic Senate (SA) at USU. She is actively involved in professional organizations such as LPPOM MUI Sumatera Utara, where she is Vice Director. She is also involved in various projects as a professional engineer, actively becoming a reviewer and assessor for ministry programs and accreditation boards, journals and reputable conferences, and involved in community service. Currently serving as Head of the Master and Doctoral Program of Chemical Engineering USU, Dr. Rosdanelli continues to excel in her primary roles as a senior lecturer and researcher in chemical engineering, specifically in the field of drying technology, pyrolysis, extraction, and bioprocesses.

Dr. Apri Heri Iswanto, SHut, MSi

Affiliation: Faculty of Forestry, Universitas Sumatera Utara.
Email: apri@usu.ac.id
Online Profile: https://fhut.usu.ac.id/id/dosen/apri-heri-iswanto

Dr. Apri Heri Iswanto was born in Pekalongan, Central Java on April 21, 1980. Since December 2003, he has served as a Lecturer in Forestry at the Faculty of Forestry, Universitas Sumatera Utara. In August 2023, he became a Professor of Forestry with expertise in articleboard and Fiberboard Technology. In addition to teaching, he is involed in scientific research activities related to wood and non-wood as raw materials for composite boards. Additionally, he is a reviewer for several reputable international journals, including: Journal of Building Engineering; Journal of Cleaner Production; Industrial Crops & Products; International Journal of Mechanics and Materials in Design; Construction and Building Materials;

Environmental Technology & Innovation; Journal of the Korean Wood Science and Technology; Case Study in Construction Materials, Biomass Conversion and Biorefinery; BioResources; Polymers; Forests; and Frontier in Materials. He is also a reviewer for USU internal research, DRTPM DIKTI research, and for the Higher Education Council of the Ministry of Education, Culture, Research and Technology (for the matching fund grant program). From 2018 to 2021, he was the Editor in Chief of the Journal of Sylva Indonesiana (Sinta-3); in 2023 he was the Editor in Chief of the Global Forest Journal. Since January 2024, Dr. Iswanto has also been an editorial board member of the Sylva Lestari Journal (Scopus Q3).

Halimatuddahliana, ST, MSc, PhD

Affiliation: Chemical Engineering at the Faculty of Engineering, Universitas Sumatera Utara
Email: halimatuddahliana@usu.ac.id
Online Profile: https://ft.usu.ac.id/id/dosen/halimatuddahliana

Dr. Halimatuddahliana was born in Medan on April 8th, 1973. She is a Professor of Chemical Engineering at the Faculty of Engineering, Universitas Sumatera Utara. She her bachelor's degree (S1) in 1997 at the Department of Chemical Engineering, Faculty of Engineering, Universitas Sumatera Utara. Since 1998, she has been part of the teaching staff at the Department of Chemical Engineering, Faculty of Engineering, Universitas Sumatera Utara.

In 2002, she completed a Master's Program (S2) followed by a Doctoral Program (S3) in Polymer Technology at Universiti Sains Malaysia in 2006. Dr. Halimatuddahliana's field of expertise is Polymer/Biopolymer-based Composite Technology. The author is a lecturer for several courses in the Department of Chemical Engineering for undergraduate, postgraduate and doctoral degrees such as Polymer Technology, Materials and Corrosion Technology, Research Methodology, Heat Transfer, Advanced Polymer Technology, Industrial Ecology, and Biopolymer Technology. In addition to being a member of the teaching staff at the Department of Chemical Engineering, she is Head of the Chemical Industrial Process Laboratory (2021-present). Dr. Halimatuddahliana actively researches, writes, participates in international seminars, and plays an active role in community service.

Dr. Hamidah Harahap, MSc

Affiliation: Chemical Engineering at the Faculty of Engineering, Universitas Sumatera Utara
Email: hamidah_usu@yahoo.com
Online Profile: https://ft.usu.ac.id/id/dosen/hamidah-harahap

Dr. Harahap, is a Professor of Chemical Engineering at the Faculty of Engineering, Universitas Sumatera Utara. She was born in Medan on October 29, 1967. She completed her bachelor's degree (S1) in 1992 at the Department of Chemical Engineering, Faculty of Engineering, Universitas Sumatera Utara. Since 1995, she has been a teaching staff member at the same department.

In 2002, she completed a Master's Program (S2) followed by a Doctoral Program (S3) in Polymer Technology at Universiti Sains Malaysia, finishing in 2010. Her field of expertise is Latex and Polymer/Biopolymer Technology. She lectures on several courses in the Department of Chemical Engineering for undergraduate, postgraduate, and doctoral degrees, including Basic Chemistry, Organic Chemistry, Chemical Industry Processes, Engineering Drawing, Polymer Technology, Research Methods, Occupational Health and Safety Engineering, and Latex Technology.

In addition to her teaching role, she has served as Deputy Director II at the Postgraduate School of Universitas Sumatera Utara since 2021. She actively engages in research, writing, and participating in international seminars, and plays an active role in community service.

Index

A

B

H

I

J

K

L

M

N

R

S

π